PHYSICAL PROCESSES IN LOW-TEMPERATURE GAS-DYNAMIC LASERS

Proceedings of the Institute of General Physics
Academy of Sciences of the USSR
Series Editor: A.M. Prokhorov

Volume 1	**Oceanic Remote Sensing,** Edited by F.V. Bunkin and K.I. Volyak.
Volume 2	**Laser-Induced Raman Spectroscopy in Crystals and Gases,** Edited by P.P. Pashinin.
Volume 3	**Magnetic and Electron Structures of Transition Metals and Alloys,** Edited by V.G. Veselago and L.I. Vinokurova.
Volume 4	**Laser Techniques for Investigation of Defects in Semiconductors and Dielectrics,** Edited by A.A. Manenkov.
Volume 5	**Fiber Optics,** Edited by Ye.M. Dianov.
Volume 6	**Nonlinear Optics and Acoustics of Fluids,** Edited by F.V. Bunkin.
Volume 7	**Formation and Control of Optical Wavefronts,** Edited by P.P. Pashinin.
Volume 8	**Lithography in Microelectronics,** Edited by T.M. Makhviladze.
Volume 9	**Selective Laser Spectroscopy of Activated Crystals and Glasses,** Edited by V.V. Osiko.
Volume 10	**Slow Combustion of Laser Plasma and Optical Discharges,** Edited by A.M. Prokhorov
Volume 11	**Resonant Heterogeneous Processes in a Laser Field,** V.A. Kravchenko, A.N. Orlov, Yu.N. Petrov, and A.M. Prokhorov
Volume 12	**Physical Processes in Low-Temperature Gas-Dynamic Lasers,** Edited by V.K. Konyukhov
Volume 13	**Effect of Laser Radiation on Absorbing Condensed Matter,** Edited by V.B. Fedorov

Proceedings of the Institute of General Physics
Academy of Sciences of the USSR
Series Editor: A.M. Prokhorov
Volume 12

PHYSICAL PROCESSES IN LOW-TEMPERATURE GAS-DYNAMIC LASERS

Edited by V.K. Konyukhov
Translated by S.A. Stewart

Nova Science Publishers
New York

Deputy Series Editor: T.B. Voliak

Nova Science Publishers, Inc.
283 Commack Road
Suite 300
Commack, New York 11725

This book is being published under exclusive English language rights granted to Nova Science Publishers, Inc. by the All-Union Copyright Agency of the USSR (VAAP).

Library of Congress Cataloging-in-Publication Data available upon request

ISBN 0-941743-90-X

The original Russian-language version of this book was published by Nauka Publishing House in 1988

Graphic Design by Elenor Kallberg and Peggy Harvey

Printed in the United States of America

CONTENTS

Part I

VIBRATIONALLY NONEQUILIBRIUM PROCESSES IN MOLECULAR LASERS

KINETICS OF HETEROGENEOUS PROCESSES IN GAS-DYNAMIC LASERS

V. N. Faizulaev

0. Introduction

In recent years, one of the promising branches in quantum electronics - gas-dynamic lasers - has advanced greatly. It encompasses a wide circle of both purely scientific and technical issues related to developing new and improving existing methods for directly converting thermal energy into laser radiation energy. The ideas forming the basis of thermal methods for exciting lasers were developed in Refs. [1-4]. The first specific proposals for making an infrared gas-dynamic laser based on a mixture of carbon dioxide gas and nitrogen was advanced in 1966 in Ref. [5]. Since that time, studies on gas-dynamic lasers began, which were in fact successfully completed in 1970 when lasing at the 00^01-10^00 transition of molecular CO_2 with a wavelength of 10.6 μm was reported almost simultaneously in a series of works [6-9].

The operation of a thermally pumped gas-dynamic laser (GDL) is based on the possibility of significantly violating equilibrium in the energy distribution between the vibrational and translational degrees of freedom of molecules during adiabatic cooling of the gas. This effect was experimentally observed for the first time when studying the gas dynamics of supersonic flow of carbon dioxide gas [10], and then, after several years, in experiments with molecular nitrogen [11]. In essence it consists of the fact that rapid cooling of a gas in supersonic flows and the concomitant sudden suppression of relaxation processes induce quenching of the molecular vibrations. The earlier the onset of quenching is, the higher the average level of vibrational excitation turns out to be and, consequently, the share of initial thermal energy is higher, which is retained by the flux in the form of a nonequilibrium reserve of vibrational quanta of the molecules. In a thermally pumped GDL this is achieved by using supersonic nozzles (or slits) ensuring high initial gas cooling rates. A supersonic nozzle fills the region of the gas-dynamic tract, where the thermal energy is coupled into the gain medium and is converted into laser radiation. Continuous operation of the GDL is thereby provided, as is high output power due to the rapid replacement of the "exhaust" portions of gas in the resonator.

A fortunate combination of the principles of thermal excitation and a moving gain medium in gas-dynamical means of producing a population inversion suggests making GDL in a discharge of the most promising gaseous coherent radiation sources. In addition to high output power, they are distinguished by simple design and autonomous power supply when the combustion products of various propellants are used as the gain medium. Nevertheless, from a series of energy characteristics such as the efficiency and specific power output, gas-dynamic lasers still noticeably lag behind cw rapid-flow gas lasers with chemical or electrical pumping. If one speaks of a CO_2 gas-dynamic laser (CO_2-GDL), the main reason for this lag is hidden in the low efficiency of coupling thermal energy into the vibration of nitrogen molecules, which are the gain medium's energy carrier.

There are several basic ways to solve the problem of improving the efficiency of a GDL. The first of these is based on using high-temperature heating in the GDL with separate thermal excitation of the gain medium in the combustion process [4, 12-15]. However, realizing the optimum operating conditions of high-temperature GDL, acting in a scheme of mixing gaseous components in a supersonic flow, is accompanied by a series of significant technical difficulties.

The second means is based on using the principle of a moving gain medium with the application of different nonthermal (chemical, electrical) excitation methods [15-19], ensuring high efficiency of coupling energy into the vibrational degrees of freedom of the molecular energy carrier. This type of GDL is closest to gas lasers with rapid through-pumping. Therefore, acquiring the distinctive properties from the latter characteristics (high efficiency and specific power output), they lose the advantages which distinguish thermally pumped GDL, namely, the capability to operate autonomously with simple technology.

The third approach is related to seeking new gain media for the usual type of GDL that do not require heating at high temperatures, since radiation would be emitted due to the thermal excitation of molecules with low energy quanta. The progress attained in this direction [20-24] permits making an optimistic prognosis regarding the potential for making a high-efficiency, low-temperature, thermally-pumped GDL which might surpass rapid-flow gas lasers in terms of energy characteristics.

The approach to the low-temperature operating regime of a GDL with nonselective thermal pumping, in addition to the tasks of seeking molecules suitable for this purpose, laser levels, etc., poses a series of qualitatively new problems that are related to gas condensation, which arises unavoidably with deep cooling of the gain medium. Among these, we should point out the following practically important problems, such as:

1) investigating heterogeneous relaxation of the vibrational energy of the molecules in gas-aerosol type systems,

2) analyzing the potential for making a biphase medium GDL;

3) estimating the effect of heterogeneous losses caused by gas condensation on the operation of low-temperature GDL.

This work is also devoted to examining the aforementioned problems as they pertain to improving the efficiency of CO_2-GDL.

First examined are the questions regarding the kinetics of a CO_2-GDL with a homogeneous active medium. The study is conducted illustrating the adiabatic expansion of different compositions of gaseous mixtures allowing for the most important channels of vibrational relaxation of the molecules. The possibility of making a GDL operating on transitions between CO_2 coupled mode levels is theoretically demonstrated.

Furthermore, questions are studied relating to the heterogeneous relaxation kinetics of molecular vibrational energy in gas-aerosol type systems. The efficiency of various mechanisms of molecular surface deactivation is estimated, and the quasi-steady regime of heterogeneous molecular relaxation is analyzed in gas-disperse systems. The potential for using aerosol particles in GDL gain media is studied.

In conclusion, a concurrent examination of condensation and heterogeneous molecular relaxation is conducted and, on this basis, a quantitative estimate of heterogeneous losses in CO_2-GDL is made. The results of the calculations illustrating the effect of water vapor condensation on the parameters of a typical CO_2-GDL are compared with experimental data.

1. Kinetics of Gas-Dynamic CO_2-Lasers

Among the important aspects of gas-dynamic laser theory is the study of the kinetics of relaxation processes in expanding supersonic flows.

Despite significant progress achieved in solving this problem as it pertains to GDL operating on molecular vibrational transitions, to date the problem of adequately describing the vibrational state of a relaxing polyatomic gas under nonsteady conditions remains very pressing. This remark pertains, in particular, to low-temperature CO_2-GDL as well, whose relaxation-kinetic scheme needs to be understood more precisely.

The first detailed description of the processes, leading to the formation of a population inversion at the 00^01-10^00 transition of CO_2 during the adiabatic expansion of a mixture containing nitrogen, was done in Refs. [25-29]. The so-called thermodynamic model [28] was used to model the vibrational relaxation of CO_2 molecules. In the framework of this model, each vibrational mode is characterized by an intrinsic distribution temperature. Later, this approach was widely used in research into relaxation processes in different composition gain media [15, 19-24], including the mixture CO_2-N_2-H_2O [30-35], which was used in a CO_2-GDL using the products of hydrocarbon fuel combustion. In the latter case, the results of the calculations turned out to be in good agreement with the experimental data [33-36].

The question of the applicability of the thermodynamic model for the symmetric longitudinal and bending vibrations of CO_2 molecules was examined in detail in Ref. [37]. As is known, a Fermi resonance is observed for these vibrational modes. It was shown that in a quasi-equilibrium state the molecular distribution of the split levels of the symmetric vibrations turns out to be similar to a Treanor distribution for anharmonic oscillators. Prior to this, the authors of Ref. [39] attempted to approximately allow for the effects of a Treanor distribution of CO_2 molecules over the lower laser levels when studying the mechanism forming the population inversion formation at the 00^01-10^0, 00^01-02^00 transitions in a CO_2-N_2-He mixture. Consequently, an approach based on the anharmonic approximation of the thermodynamic model of CO_2 was used in Refs. [40-42][1] to evaluate the possibility of making a gas-dynamic CO_2-laser operating on the transition between the 03^10-10^00 levels.

This section examines the kinetics of the formation of the population inversion on the vibrational levels of molecular CO_2 in freely expanding supersonic flows. The examination is conducted in the framework of the thermodynamic model of CO_2 allowing for the fine structure of the levels of the symmetric mode. Based on the derived calculation technique, we study the physical features and characteristics of homogeneous CO_2-GDL,

---

[1] The suggestion regarding the creation of this type of GDL was examined independently in these works and was realized in Ref. [42].

operating both on the 00^01-10^00 transition using CO_2-N_2-H_2O, He mixtures and on the 03^10-10^00 transition using a CO_2-Ar mixture.

1.1. Thermodynamic Model of the Vibrational Distribution of CO_2 Molecules

There are several basic ways to approximately describe the vibrational state of polyatomic molecules under nonsteady conditions. The simplest idea for this, but the most complex to realize, is the approach based on numerically solving the balance equations for the populations of the individual levels. This approach usually requires detailed knowledge of the kinetics of the population of a large number of levels, which in many cases cannot be done in practice. Therefore, the quasi-equilibrium vibrational distribution method has become widely used at the present time. It permits one to significantly simplify solving the multilevel problem and to reduce it to determining the time dependence of a comparatively small number of energy parameters of a thermodynamically nonequilibrium vibrational system.

A classical example of this approach is the description of the vibrational relaxation of polyatomic molecules using the thermodynamic model [28], in which it assumed that the rates of vibrational exchange inside each ν_m-mode of vibration (*VV*-process)

$$v_m, v_m - p_m \xrightarrow{R_{VV}} v_m + p_m, v_m \tag{1}$$

significantly exceeds the rate at which quanta are exchanged between different $m \neq n$ vibrational modes $R_{VV'}$ (*VV'*-process)

$$v_m, v_n - p_n \xrightarrow{R_{VV'}} v_m + p_m, v_n \tag{2}$$

and vibrational-translational relaxation R_{VT} (*VT*-process)

$$v_m \xrightarrow{R_{VT}} v_m - p_m. \tag{3}$$

Here, v_m is the index of the molecular vibrational level and p_m is the number of quanta lost or gained by the ν_m-oscillator during the transition. In this case, the vibrational relaxation of polyatomic molecules proceeds in several stages that have significantly different time scales.

First, intermode quasi-equilibrium is rapidly established (after a time on the order of $\tau \sim 1/R_{VV}$). In the harmonic approximation it is described by a Boltzmann distribution function of the molecules relative to the levels of each ν_m vibrational mode

$$W_{v_m} \sim g_{v_m} \exp(-v_m \omega_m / T_m) \tag{4}$$

with a vibrational temperature T_m determined by the intrinsic quantum reserve

$$e_m = r_m [\exp(\omega_m/T_m) - 1]^{-1}, \tag{5}$$

where ω_m, r_m are the energy of the quantum and the multiplicity of the degeneracy of the ν_m vibration, respectively; g_{v_m} is the statistical weight of the v_m-level. In the final stage, a thermally equilibrium vibrational distribution of the polyatomic molecules with the gas temperature T is established. If selective VV'-exchange is possible in the vibrational system, i.e., occurring intensely only over one of the channels of Eq. (2), then for $R_{VV'} >> R_{VT}$ this stage will precede the establishment after time $\tau \sim 1/R_{VV'}$ of the intermode quasi-equilibrium with the vibrational temperatures T_m and T_n [43]:

$$p_m \omega_m / T_m - p_n \omega_n / T_n = (p_m \omega_m - p_n \omega_n)/T. \tag{6}$$

In the vibrational system of polyatomic molecules, such a distribution is usually realized for those types of oscillations whose frequency approximately satisfies the resonance condition $p_m\omega_m = p_n\omega_n$, as in this case it is known that intermode exchange by quanta should proceed quite rapidly.

For CO_2 molecules, whose energy level diagram is shown in Fig. 1, the resonance condition is satisfied for longitudinal ν_1 and bending ν_2 symmetric vibrations: $\omega_1 = 2\omega_2$. In connection with this, the thermodynamic model of CO_2 is described by the distribution

$$W_{v_1 v_2 v_3} \sim \prod_{m=1}^{3} g_{v_m} \exp(-v_m \omega_m / T_m), \tag{7}$$

in which the temperatures of the symmetric vibrations T_1 and T_2 are assumed to be equal [15, 19].

Although this assumption is obvious in the harmonic approximation, it needs to be refined in the anharmonic approximation. The crux of the matter is that due to the Fermi resonance, the symmetric vibrations turn out to be strongly connected (coupled), and their randomly degenerate energy levels in the harmonic approximation turn out to be split by components of the corresponding multiplets (see Fig. 1). For this reason the vibrations ν_1 and ν_2 can no longer be examined independently, ascribing intrinsic vibrational temperatures to them. In such a situation, for these vibrations it is judicious to assign a single effective vibrational temperature T_{12} of the Treanor distribution of the energy levels of the symmetric modes according to the relation [37]

$$W_{v_1 v_2^l 0} \sim g_{v_1 v_2^l 0} \exp\left(-v_{12}\,\omega_2/T_{12} - \Delta E_{v_1 v_2^l 0}/T\right), \tag{8}$$

where $\Delta E_{v_1 v_2^l 0} = E_{v_1 v_2^l 0} - v_{12}\omega_2$ is the shift in the energy level $v_1 v_2^l 0$ relative to the center of the multiplet with vibrational number $v_{12} = 2v_1 + v_2$; l is the vibrational momentum of oscillator ν_2.

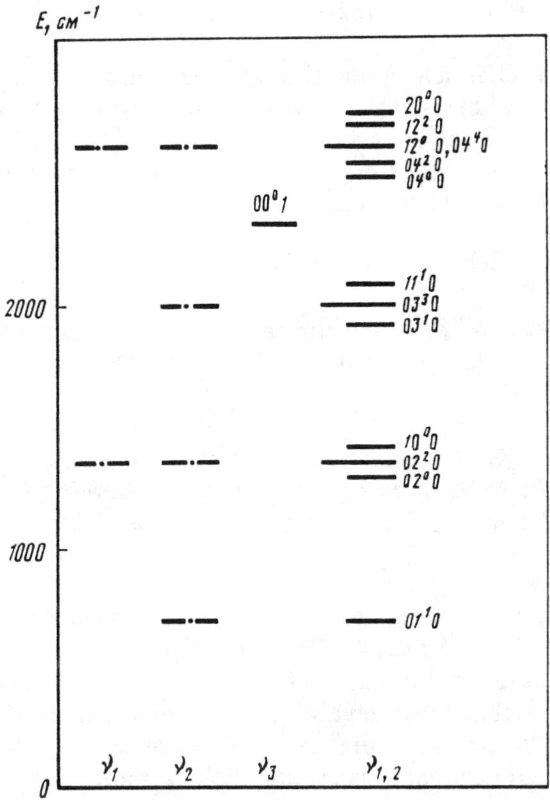

Fig. 1. Diagram of the lower vibrational levels of the CO_2 molecule. ν_1, ν_2 are the levels of the symmetric longitudinal and bending vibrations in the harmonic approximation, ν_{12} is in the anharmonic approximation, and ν_3 is the antisymmetric mode.

This distribution can be treated as an anharmonic approximation of the thermodynamic model for the symmetric vibrations of CO_2. It is formed as a result of rapid VV'-exchange and transitions inside the multiplets of the symmetric mode and is a consequence of conservation of the supply of quasi-particles in selective processes of the type in Eq. (2)

$$V = v_m/p_m + v_n/p_n. \tag{9}$$

The role of the latter in this case is played by either p_m quanta of oscillators ν_m or p_n quanta of oscillator ν_n. The existence of the invariant in Eq. (9) permits us to directly derive the general form of the energy distribution function of a system of oscillators in terms of the energy and number of quasi-particles [43]

$$W_{V\beta} \sim g_{V\beta} \exp\left[(\mu V - E_{V\beta})/T\right], \tag{10}$$

where μ is the chemical potential of the quasi-particles, $E_{V\beta}$ is the energy of the β-th level at a given V. For resonant exchange of quasi-particles between harmonic oscillators ν_m and ν_2 this distribution should reduce to a Boltzmann distribution with a single temperature T_{mn} for the oscillators. Therefore, in our case

$$\mu = \omega_1 (T_{12} - T)/T, \quad V = v_1 + v_2/2.$$

From this it follows that the presence of fine structure of the CO_2 coupled mode levels ($E_{V\beta} \neq \omega_1 V$) leads to Eq. (8).

1.2. Technique for Modeling the Kinetics of the Population of the Vibrational Levels of CO_2 Molecules in the Gain Media of CO_2 Gas-Dynamic Lasers

In the preceding subsection, the basic approximations that are used when analyzing quasi-resonant vibrational distributions for carbon dioxide gas molecules were examined. The question of allowing for the fine structure of the vibrational levels of the symmetric mode turned out to be significant. Fine structure is also important when describing the kinetics of the vibrational relaxation of CO_2 molecules that occur when symmetric vibrations take part. However, the analytical difficulties arising here turn out to be significantly greater. This is because, when deriving the *VT*- and *VV'*-relaxation equations for the symmetric mode, it is necessary to consider not only the specifics of the quasi-equilibrium molecular distribution of multiplet levels, but also the difference of the adiabatic factors in the vibrational transition probabilities. It is not possible to analytically solve this problem in general form. It can only be hoped that (and the calculations done in Ref. [37] are partially convincing in this), when using experimental data on the rates of the corresponding processes, the harmonic approximation of the thermodynamic model is completely suitable for describing the vibrational relaxation of CO_2 molecules.

In connection with this, it appears useful to calculate in the following manner the kinetics of the population of the CO_2 vibrational levels under nonsteady conditions corresponding to supersonic gas flow. In the first stage using the well-known vibrational relaxation equations [44, 45], derived in the framework of the harmonic approximation of the thermodynamic model of CO_2, we determine the average supply of quanta e_m for each ν_m-mode individually, and then using these data we find the population distribution $N_{v_1 v_2^{l} 0}$ of the vibrational levels corresponding to the anharmonic approximation of the same model. In this approach it is assumed that the process of establishing a Treanor distribution of the symmetric mode levels takes place at a rate $R_{VV'}$ which significantly exceeds the rate that the gas cools, or, more precisely,

$$(\Delta E_{12}/T)\,|\,d\ln T/dt\,| \ll R_{VV'},$$

where ΔE_{12} is the average amount of the multiplet splitting energy of the symmetric mode. Calculations show that this condition is satisfied in the majority of cases of practical interest for CO_2-GDL. The proposed calculation technique will be examined below in detail to illustrate describing vibrational relaxation in a CO_2-N_2 mixture.

Following the familiar classification given in the survey of Ref. [46], it is possible to distinguish the following basic channels of vibrational molecular relaxation in VT- and VV'-processes of energy exchange. These processes determine the kinetics of the population of the vibrational levels in a mixture of CO_2-N_2:

$$N_2^*(\nu_4) + CO_2 \xrightarrow{k_{43}} N_2 + CO_2^*(\nu_3), \tag{11}$$

$$CO_2^*(\nu_3) + M_i \xrightarrow{k_{32}} CO_2^*(\nu_1\nu_2) + M_i, \tag{12}$$

$$CO_2^*(\nu_2) + M_i \xrightarrow{k_{20}} CO_2 + M_i. \tag{13}$$

The asterisk denotes the vibrationally excited state of the molecule and which vibrational states are excited are given in parentheses.

In the harmonic approximation, the energy relaxation kinetics of the $CO_2(\nu_1\nu_2\nu_3)$ modes and vibrations of $N_2(\nu_4)$ are described by the following system of equations [44]:

$$de_4/dt = -k_{43}\,p^{(1)}(e_4 - e_3) - \sum_i k_{40}^{(i)}\,p^{(i)}(e_4 - \bar{e}_4), \tag{14}$$

$$de_3/dt = -k_{43}\, p^{(2)}(e_3 - e_4) - \sum_i \beta k_{32}^{(i)} p^{(i)}(e_3 - \tilde{e}_3)\,, \tag{15}$$

$$de_{12}/dt = 3 \sum_i \beta k_{32}^{(i)} p^{(i)}(e_3 - \tilde{e}_3) - \sum_i k_{20}^{(i)} p^{(i)}(e_2 - \bar{e}_2), \tag{16}$$

where

$$\beta = \left(\frac{2 + e_2}{2 + \bar{e}_2}\right)^3 ; \quad \tilde{e}_3 = (1 + e_3)\left(\frac{e_2}{2 + e_2}\right)^3 \exp\left(-\frac{500}{T}\right);$$

$e_{12} = e + 2e_1$ is the energy of the symmetric mode of CO_2 normalized to the magnitude of quantum of the bending vibration, $\bar{e}_m$ is the value of e_m at $T_m = T$, $k_{mn}^{(i)}$ is the rate constant of the $\nu_m \to \nu_n$ process in collisions with molecules M of the ith sort, and $p^{(i)}$ is the pressure of the ith component of the mixture ($i = 1$ for CO_2, $i = 2$ for N_2).

The first terms in Eqs. (14) and (15) describe the resonant transfer of the vibration with the exchange of vibrational quanta between $N_2(\nu_4)$ and the antisymmetric vibrational mode of $CO_2(\nu_3)$, the second term in Eq. (15) and the first term in Eq. (16) describe nonresonant vibrational exchange over the most probable reaction channel $\nu_3 \to 3\nu_2 + \Delta E$ with $\Delta E = 500$ K, and the remaining terms in Eqs. (14) and (16) describe vibrational-translational relaxation of the corresponding vibrations. When writing Eq. (16) it was assumed that the relaxation of the longitudinal and symmetric bending vibrations of CO_2 take place simultaneously, such that $T_1 = T_2$. In addition, the term in Eqs. (15) and (16) corresponding to reaction (12) is written in a form which permits one to directly use the experimental values of the rate constants of the VV'-relaxation of the antisymmetric vibrational mode. It is easy to see that for $T_{1,2} = T$, when $\beta = 1$, $\tilde{e}_3 \simeq \bar{e}_3$, it essentially describes the usual process of VT-relaxation of the ν_3 vibration.

To completely describe the kinetics of the vibrational relaxation of molecules in a supersonic flow of an expanding gas, system (14)-(16) should be supplemented with the corresponding gas-dynamic equations, in which the effects of heat liberated from the vibrational degrees of freedom are taken into account. In some cases, however, it turns out to be possible to treat the dynamics of a vibrationally nonequilibrium gas in the approximation of adiabatic motion of a medium with an effective adiabatic index γ, corresponding to a certain degree of quenching of the vibrational component of the heat capacity of the gas. This significantly simplifies the calculation, since it permits one to use the well-known solution of the gas-dynamic equations for isentropic flows.

In the final stage of the calculation, the current values of the vibrational temperatures $T_1 = T_2 \equiv T_{12}$ and T_3 that were found by solving Eqs. (14)-(16) for both modes of CO_2 are taken as the initial parameters for the quasi-equilibrium population distribution:

$$N_{v_1 v_2^l v_3} = N_{000}\, g_l \prod_{m=1}^{3} (x_m)^{v_m}\, \varphi_{v_1 v_2^l 0}, \tag{17}$$

$$N_{000} = n \prod_{m=1}^{3} (1 - x_m)^{r_m},$$

where $x_m = \exp(-\omega_m/T_m)$, $\varphi_{v_1 v_2^l 0} = \exp(-\Delta E_{v_1 v_2^l 0}/T)$, $g_l = 1$ if $l = 0$, $g_l = 2$ if $l \neq 0$, and n is the concentration of CO_2 molecules.

1.3. Formation of the Population Inversion at the 00^01-10^00 Transition of CO_2 Molecules in a Supersonic Jet of Diverging Gas

We shall turn now to studying the kinetics of CO_2-GDL with nonselective thermal pumping of the gain medium, in which nitrogen acts as the energy carrier for the carbon dioxide gas radiating at the transition 00^01-10^00, and water vapor or helium acts as an auxiliary component serving to vacate the lower laser level. We shall conduct the examination on an example of gas flowing from a slit into a vacuum, using the technique described above for calculating the populations of the vibrational levels of CO_2 molecules.

The system of Eqs. (14)-(16) for the vibrational relaxation of CO_2 and N_2 molecules in CO_2-N_2-H_2O, He mixtures was solved numerically on a computer using the Runge-Kutta method. The dependence of the gas-dynamic parameters on the distance to the slit ξ was assigned for the central line of flow of the jet starting from the solution of the gas-dynamic equations for planar stable isentropic gas flow from the slit into the vacuum [47]. The corresponding calculation by the method of characteristics was done by E'. A. Ashratovy and G. K. Bunina. Table 1 gives the results of the calculation of the Mach number M along the jet for different values of γ. The distance from the slit $\xi = x/h_0$ is expressed in calibers, where h_0 is the half-width of the slit.

Table 1

Mach Number Values on the Axis of the Jet

γ	ξ						
	1	2	5	10	20	50	100
1.3	1.48	2.0	2.87	3.58	4.2	5.15	5.9
1.4	1.5	2.06	3.0	3.82	4.68	5.95	7.0
1.5	1.51	2.1	3.2	4.16	5.23	6.9	8.37

The experimental values of the rate constants of processes (11)-(13) were taken in correspondence with their approximation functions cited in Ref. [45]. The state of the gas right at the slit, i.e., for $\xi = 0$, is assumed to be in thermal equilibrium with a temperature and pressure determined by the relations

$$\frac{T}{T_s} = \left(1 + \frac{\gamma - 1}{2} M^2\right)^{-1}, \quad \frac{p}{p_s} = \left(1 + \frac{\gamma - 1}{2} M^2\right)^{-\gamma/(\gamma-1)} \tag{18}$$

for $M = 1$. The gas in front of the slit is assumed to be stationary at stagnation temperature T_s and pressure p_s. The effective adiabatic index is calculated from the formula

$$\gamma = 1 + 2/(5 + 4\kappa_1 + \kappa_3 - 2\kappa_4), \tag{19}$$

where κ_i is the partial fraction of the ith component in the mixture ($\Sigma\kappa_i = 1$, $i = 1$ corresponds to CO_2, 2 - N_2, 3 - H_2O, 4 - He). Allowed for here, by virtue of the quenching effect and due to the insufficient excitation of vibrations, is the fact that it is possible to neglect the contributions to the vibrational component of the heat capacity of N_2 and H_2O in the gaseous mixture, and also the antisymmetric and symmetric longitudinal vibrations of CO_2.

The vibrational-rotational population inversion

$$\Delta N_{vJ} = N'_{vJ} - (g'_{vJ}/g_{vJ}) N_{vJ}$$

was calculated for the $P(20)$ branch of the $(00^01)'$-(10^00) band in the approximation that the CO_2 molecules' distribution of rotational levels is in thermal equilibrium:

$$N_{vJ} = g_{vJ} \frac{2B}{T} N_v \exp\left(- \frac{BJ(J+1)}{T}\right), \tag{20}$$

where g_{vJ} is the statistical weight of the vibrational-rotational energy level and B is a rotational constant of the molecule. The gain index is calculated for the center of the spectral line from the equation [48]

$$\alpha = \frac{\lambda^2}{8\pi^2} \frac{A_{21}}{\Delta\nu_c} \Delta N_{vJ} S(a), \tag{21}$$

in which the concurrent action of Doppler and collisional broadening mechanisms is taken into account:

$$S(a) = \sqrt{\pi}\, a \exp(a^2)\, [1 - \mathrm{erf}\,(a)],$$

where $a = \Delta\nu_c \sqrt{\ln 2} / \Delta\nu_D$ and $\Delta\nu_D$ and $\Delta\nu_c$ are the Doppler and collisional line-widths, respectively. The probability of a radiative transition A_{12} and data on the collisional broadening cross-sections in the collisions of CO_2 molecules with N_2, He, and CO_2 molecules were taken from Ref. [49]. The broadening cross-section of CO_2 in water vapor was assumed to be the same as in carbon dioxide gas. The parameter $h_0 = 0.04$ cm.

Figure 2 shows typical vibrational T_v and gas T temperature distributions along the axis of the jet that are typical for a CO_2-N_2-He mixture. It is evident that immediately behind the slit the N_2 vibrations are quenched and those directly in front are found to be at the resonance of the antisymmetric vibrational mode of CO_2. Here, the temperature of the symmetric vibrations as a consequence of the higher rate of VT-relaxation, caused chiefly by the presence of helium in the mixture, continues to drop and at distances of only 2-3 calibers from the slit begins to break away from the gas temperature. Thus, a level of cooling of the symmetric vibrations is also provided at which the formation of a population inversion at the 00^01-10^00, 02^00 transitions is possible. In the case of the 00^01-10^00 transition, this also corresponds to the effect of a Treanor distribution of CO_2 molecules relative to the components of the symmetric mode multiplets, as a result of which a distinct additionally lower temperature of the lower laser level population is possible at low gas temperatures ($T < 150$ K) (see Fig. 2). It is obvious that for the 00^01-02^00 laser transition competing with it, this mechanism of the depopulation of the levels should inhibit the formation of a population inversion.

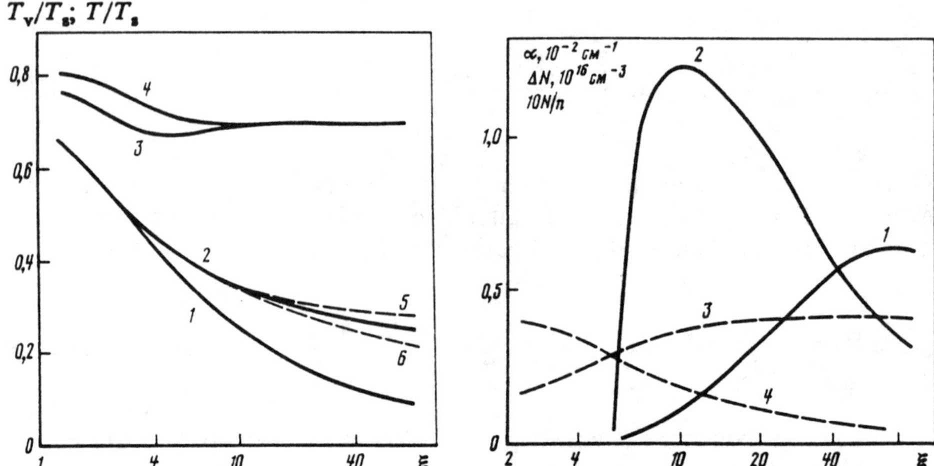

Fig. 2. Distribution of vibrational and gas temperatures along the jet. Mixture of CO_2:N_2:He = 1:4:4:, T_s = 1800 K, p_s = 54 atm. *1*) T/T_s, *2*) T_2/T_s, *3*) T_3/T_s, *4*) T_4/T_s, *5, 6*) temperatures of the population of the 02^00 and 10^00 levels, respectively.

Fig. 3. Distribution of the gain index (*1*) and vibrational population inversion (*2*) for the 00^01-10^00 transition along the jet axis. Mixture of CO_2:N_2:He = 1:4:4:, T_s = 1800 K, p_s = 54 atm. *3, 4*) relative populations of the 00^01 and 10^00 levels, respectively.

Figure 3 shows the axial distributions of the relative populations of the vibrational levels $N_{v_1 v_2^l v_3}/n$, the gain index α, and the population inversion for the 00^01-10^00 transition in a CO_2-N_2-He mixture. The variation of the population inversion with distance is caused by the competition of two basic factors: the increase in the discontinuity between the temperatures of the population of the upper and lower laser levels and the decrease in the gas density. Typical for this variation is a distinct maximum which exists at a small distance from the slit and is situated right on the edges of the region where the population inversion exists in the jet. The effect of enriching the fundamental vibrational state of the CO_2 molecules enhances this to some degree. This is caused by the cooling of the symmetric vibrations. In this case, it not only compensates for the existing small drop in the temperature of the antisymmetric vibrations, but also leads to an increase in the relative population of the 00^01 level (see Fig. 3). In distinction from the population inversion, the maximum gain index turns out to be mildly sloping and located far away from the slit. Simple analysis of the variation of α with ΔN_v and the form factor of the *S* line shows that the maximum α_{max} should be at distances where Doppler and collisional line broadening are roughly comparable in magnitude.

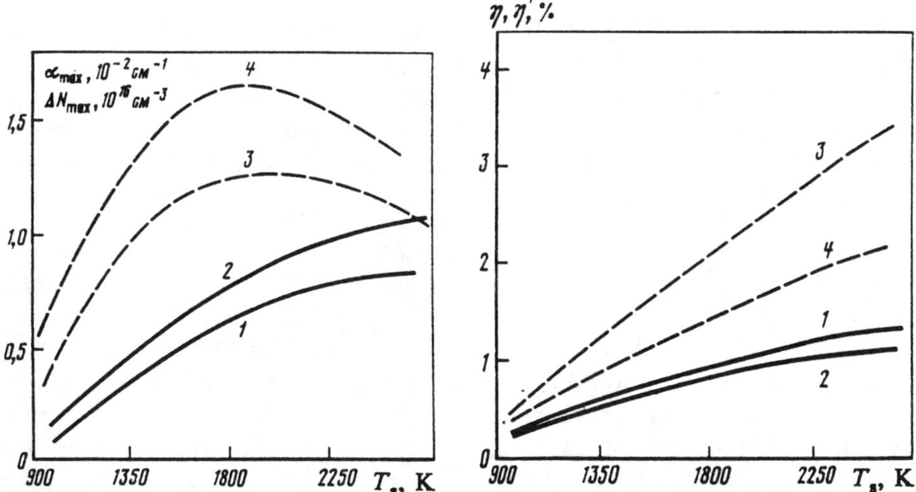

Fig. 4. Variation of the maximum gain index (*1, 2*) and vibrational population inversion (*3, 4*) at the 00^01-10^00 transition as a function of the stagnation temperature.
1, 4) mixture of CO_2:N_2:He = 1:4:4; *2, 3*) mixture of CO_2:N_2:H_2O = 1:8:0.5; initial gas density $n_s = 2.2 \cdot 10^{20}$ cm^{-3}.

Fig. 5. Variation of the thermal pumping efficiency in a CO_2-GDL at the 00^01-10^00 transition (*1, 2*) and its limiting value (*3, 4*) as a function of stagnation temperature.
1, 4) mixture of CO_2:N_2:He = 1:4:4; *2, 3*) mixture of CO_2:N_2:H_2O = 1:8:0.5; $n_s = 2.2 \cdot 10^{20}$ cm^{-3}.

Figure 4 shows the variation of α_{max} and ΔN_{max} with stagnation temperature. For comparison, the results of the calculation referring to the CO_2-N_2-He mixture are shown along with the corresponding data for the CO_2-N_2-H_2O mixture. The later has obtained wide application in CO_2-GDL because the products of burning hydrocarbon fuels in air can be utilized. The maximum at $T_s \approx 1800$-2000 K is typical for the temperature dependence of the population inversion, whereas the maximum in the gain coefficient occurs at $T_s \approx 2700$-3000 K. This feature is explained by the fact that with an increase in T_s, along with an increase in the initial reserve of vibrationally excited CO_2 and N_2 molecules, the effective rate of their concurrent vibrational and translational relaxation processes described by Eqs. (11) and (12) also increases. While the first process promotes an increase in α and ΔN_v with increasing T_s, the second process obviously inhibits it.

The same can be said for the temperature dependence of the efficiency of nonselective thermal pumping in a CO_2-GDL (Fig. 5). Its value η is determined by the ratio of the part of the vibrational energy of N_2 and CO_2 molecules ϵ, which can be converted to laser radiation in an ideal resonator, to the thermal energy spent in the preliminary heating of the gas mixture:

$$\eta = \eta_0(\epsilon - \epsilon^*)/c_p T_H, \qquad \epsilon = \omega_4(\kappa_2 e_4 + \kappa_1 e_3), \tag{22}$$

where ϵ^* is the minimum value of ϵ achieved with the condition of there being a population inversion ($T_3 \geq 1.7T_2$), $\eta_0 = 0.41$ is the quantum efficiency of the laser transition, c_p is the specific heat of the gas mixture heated at constant pressure to the temperature T_s. From Fig. 5 it is clear that the thermal pumping efficiency even at high stagnation temperatures ($T_s \approx$ 2400-2800 K) barely exceeds 1%. Such low values of η for both mixtures are caused not only by the high magnitude of the quantum of the nitrogen molecules (ω_4 = 3350 K), but also, as was mentioned, by relaxation losses of the vibrational energy of the N_2 molecules through the carbon dioxide additive.

The effect of the latter factor can be calculated by comparing the value of η with the limiting value of the thermal pumping efficiency η' shown in Fig. 5, which corresponds to the instantaneous quenching of the vibrations of N_2 molecules. By comparing these data it follows that the relaxation processes can significantly lower the efficiency of using high-temperature heating of the mixtures in CO_2-GDL. In this respect, the advantages of high-temperature CO_2-GDL with selective thermal pumping are obvious. Here, only slowly relaxing nitrogen is subjected to preheating and the required composition of the gain medium is formed in the process of mixing the gases directly in the supersonic flow of vibrationally nonequilibrium N_2 [12-15]. As for low-temperature CO_2-GDL, the problem of improving the thermal pumping efficiency requires a more radical solution. One of the most promising solutions for this will be examined in the next subsection.

1.4. Gas-Dynamic Laser Operating at Transitions Between Coupled Mode Levels of CO_2

The possibility of obtaining amplified radiation in the 18.4 μm region at the vibrational-rotational transitions of the 03^10-10^00 CO_2 band was first indicated in Ref. [37]. It was suggested that the population inversion at this transition arises as a consequence of the components of coupled mode multiplets having a Treanor molecular distribution. The corresponding calculations were performed in Ref. [50] for the parameters of the gain medium for a flow-through variant of a CO_2 gas-dynamic laser, in which the low-temperature (T = 150 K) operating regime was ensured by argon acting as a heat carrier. However, the gain coefficient at the 03^10-10^00 transition turned out to be low due to the need for the carbon dioxide gas to be strongly diluted with argon (CO_2:Ar = 1:50). In conjunction with this, the potential for using thermal pumping in a gas-dynamic variant of a CO_2-laser operating on the coupled-mode levels is of interest.

As is known, for there to be a population inversion with the molecules having energy levels in a Treanor distribution, considerable vibrational nonequilibrium under low gas temperature conditions is required. In this case, the nonequilibrium of the CO_2 molecule's energy level distribution of the symmetric mode should be such that the following relation is satisfied for the vibrational temperature T_{12} and gas temperature T:

$$T_{12}/T \geqslant [1 - (E_{v+1,\lambda} - E_{v,\beta})/\omega_2]^{-1}. \qquad (23)$$

For example, for the 03^10 and 10^00 levels, which pertain to multiplets with vibrational quantum numbers v = 3, v = 2 and are shifted relative to their centers by the energy $\Delta E_{v_1 v_2^l 0}$, and equal -70 and 50 cm^{-1}, respectively, according to Eq. (23) it is possible to have a population inversion only for $T_{12} \geq T_{12}^* = 5.4T$.

In essence, Eq. (23) is the analog to the partial population inversion condition on the vibrational-rotational transitions of diatomic molecules [4], for which the rotational levels act as the multiplet components. For the gas-dynamic means of thermally pumping the gain medium, the feasibility of Eq. (23) is ensured by the effect of quenching the symmetric vibrations that emerges in the process of the adiabatic expansion of the gas through the supersonic nozzle or slit. To reach the necessary difference between the vibrational T_{12} and gas T temperatures in the gas mixture, it is useful to introduce a relatively small amount of argon ([Ar] ~ 50%), which, without really affecting the kinetics of the relaxation processes, makes it possible to significantly increase the rate and depth of cooling of the carbon dioxide gas in a supersonic flow.

Cited below are the results of numerically solving the kinetics equations (15) and (16) that describe, in the harmonic approximation of the thermodynamic model of CO_2, the vibrational relaxation in a mixture of CO_2:Ar = 1:1 expanding through a slit into a vacuum. The gas mixture flow is assumed to be isentropic with an effective adiabatic index γ corresponding to the instantaneous quenching of all CO_2 vibrations. For the rate constants of the relaxation processes, the data cited in Refs. [51, 52] were used. The vibrational population inversion ΔN_v and the gain index α at the 03^10-10^00 transition were calculated in the approximation of the CO_2 molecules having a Treanor distribution of coupled mode energy levels. The temperature of the rotational distribution of CO_2 molecules was taken to be equal to the gas temperature, and data on the collisional broadening cross-sections and the radiative transition probability are the same as in Ref. [50].

Figure 6 shows gas T and vibrational temperature distributions of symmetric T_{12} and antisymmetric T_3 modes of CO_2 along the jet axis that are typical for a CO_2-Ar mixture. Likewise shown are the temperature distributions of the populations of the 03^10 and 10^00 levels. The distance from the slit ξ is given in calibers with h_0 = 0.04 cm. It is clear from Fig. 6 that, as a result of Treanor pumping, the temperature of the upper laser level population increases rather quickly and the lower level decreases, both in proportion to distance from the slit. This also leads to the formation of a population inversion at the 03^10-10^00 transition at some distance from the slit. Note that the high level of quenching of the symmetric vibrations and the low gas temperature that are typical for a CO_2-Ar gas mixture expanding through a slit, in principle, permit the formation of a population inversion at other transitions as well. In particular, for $T_{12}/T \geq 12$ it becomes possible at the 02^00-10^00 transition in the $\lambda \approx 16$ μm region. However, this can be prevented by condensation of carbon dioxide gas in the jet due to extreme cooling.

Figure 7 shows the variation of the relative populations $N_{v_1 v_2^{l}0}/n$, the gain index α, and the population inversion ΔN_v for the 03^10-10^00 transition as functions of the distance ξ to the slit. They are very similar in form to the functions corresponding to the 00^01-10^00 transition (see Fig. 3). The difference in the mechanisms forming the population inversion in this case appears primarily from the quantitative side, and namely at the more distant location of the region where there is gain at the 03^10-10^00 transition and in the absence of distinct maxima in the distributions of α and ΔN_v along the axis of the jet. As for the significantly larger values of α than in the usual type of CO_2-GDL, this in particular is related to the large relative content of carbon dioxide gas in the mixture.

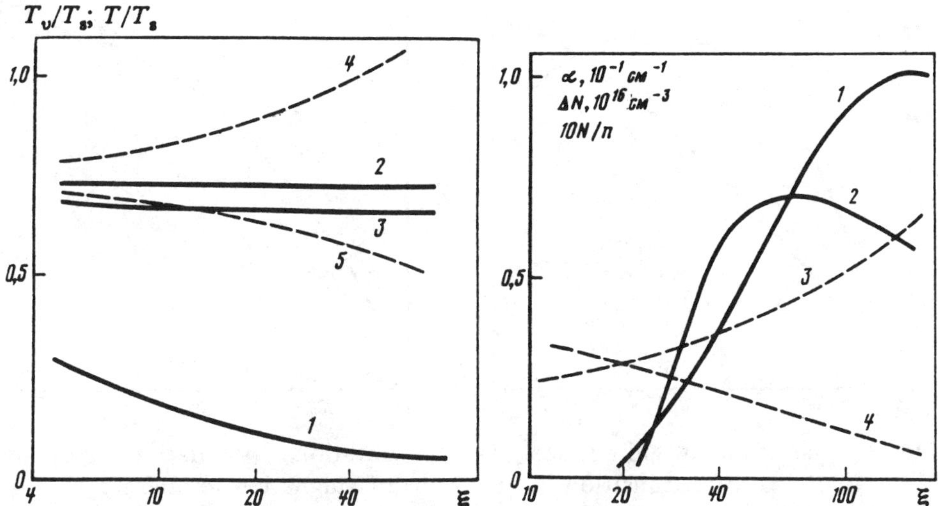

Fig. 6. Distribution of vibrational and gas temperatures along the axis of the jet.
CO_2:Ar = 1:1 mixture, T_s = 1300 K, p_s = 10 atm. *1*) T/T_s, *2*) T_2/T_s, *3*) T_3/T_s, *4, 5*) temperatures of the populations of the 03^10 and 10^00 levels, respectively.

Fig. 7. Distribution of the gain index (*1*) and vibrational population inversion (*2*) for the 03^10-10^00 transition along the axis of the jet.
CO_2:Ar = 1:1 mixture, T_s = 1300 K, p_s = 10 atm. *3, 4*) relative populations of the 03^10 and 10^00 levels, respectively.

Figure 8 shows the variation of the vibrational population inversion ΔN_v and gain index α at the 03^10-10^00 transition as functions of the stagnation temperature T_s. The values of ΔN_v and α correspond to a fixed gas temperature in the jet of T = 100 K, when condensation can be neglected when defining the initial conditions. The tendency to drop that appears for $T_s \approx$ 1600-2000 K is typical for both functions. This is related to the increase in the *VT*-relaxation rate, and also to the highly excited vibrational states of CO_2's symmetric mode being gradually populated as the stagnation temperature increases.

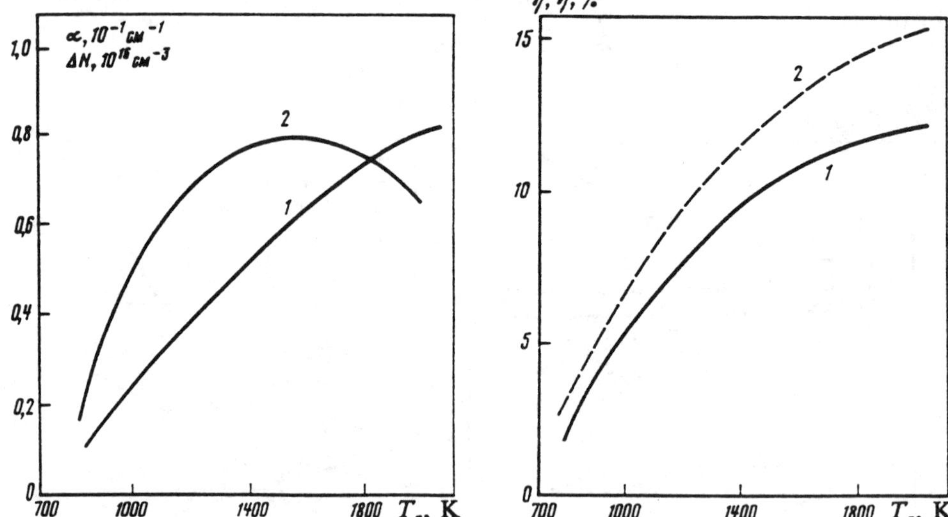

Fig. 8. Variation of gain index (*1*) and vibrational population inversion (*2*) at the 03^10-10^00 transition as functions of stagnation temperature. CO_2:Ar = 1:1 mixture, $n_s = 0.72 \cdot 10^{20}$ cm^{-3}.

Fig. 9. Variation of the thermal pumping efficiency in a CO_2-GDL at the 03^10-10^00 transition (*1*) and its limiting value (2) as functions of stagnation temperature.
CO_2:Ar = 1:1 mixture, $n_s = 0.72 \cdot 10^{20}$ cm^{-3}.

In conclusion, we shall calculate the possible thermal pumping efficiency in a gas-dynamic CO_2-laser operating at the 03^10-10^00 transition. Assuming that the emission of laser radiation in this case is characterized by a quantum efficiency equal to unity, and is possible only under condition (23), we get

$$\eta = (\epsilon - \epsilon^*)/c_p T_s, \qquad \epsilon = \omega_2 \kappa_1 e_{12}. \tag{24}$$

The value of ϵ^* for $T_{12}^* = 5.4T$ determines the maximum expenditure of the nonequilibrium supply of symmetric vibration quanta that is achievable by the 03^10-10^00 transition. As is clear from Fig. 9, which shows the variation of $\eta(T_s)$, the thermal pumping efficiency in a gas-dynamic CO_2 laser at this transition can be 3-12% for T_s = 800-1500 K, respectively. Such high values of η as compared to the usual type CO_2-laser at low stagnation temperatures are caused by the low energy of the vibrational quantum (ω_2 = 960 K) and the large specific heat capacity of the symmetric vibrations of carbon dioxide gas molecules.

2. Heterogeneous Molecular Relaxation and Biphase Media Lasers

In connection with the problem of producing low-temperature GDL, the study of the relaxation processes in vibrationally nonequilibrium gas-disperse systems is of great practical importance. The vibrational relaxation of molecules in such systems is complex, since it includes *VT*- and *VV'*-processes that take place with the participation of adsorbed molecules. The kinetics of these processes have been little studied so far, caused to a significant degree by the lack of information on the physical surface properties that determine the interrelationship of adsorption and vibrational relaxation of molecules. Therefore, at the present time, a very pressing task is that associated with analyzing the general relationships of heterogeneous molecular relaxation in disperse systems and to developing models for the surface deactivation mechanisms of gas molecules.

The investigations into heterogeneous molecular relaxation were initiated in Refs. [53, 54]. Although some time has passed since then, nevertheless, these studies have been conducted intensely only recently, when molecular gas lasers came on the scene and, with them, the need to account for the loss of vibrationally excited molecules at the walls. Experimental studies into the accommodation processes of the vibrational energy of CO_2, N_2, and N_2O molecules have been particularly widely developed during these years [55-59]. As for theory, it has developed in the direction of a detailed study of the effect of transport processes on the heterogeneous molecular relaxation in a gas-wall type system [60, 61]. Work in this direction was stimulated by different laser applications, and also by the problem of heterogeneous isotope separation [62, 63].

A new stage in the research into heterogeneous molecular relaxation is related to GDL, or more precisely, with the problem of accounting for heterogeneous losses that are caused by gas condensation in low-temperature GDL and with the transition to biphase gain media [64-67]. The potential for significantly accelerating vibrational molecular relaxation as a result of gas condensation was first indicated in Ref. [68]. Later, the role of aerosol particles as a catalyst for the relaxation processes was examined in detail in Refs. [67, 69]. A general analysis of vibrational relaxation of molecules in heterogeneous mixtures was also given there, and promising approaches for making GDL based on biphase media were also examined.

In this section, the basic mechanisms of surface deactivation and the molecular kinetics of vibrational relaxation in biphase gas-aerosol type systems are theoretically studied. The treatment is conducted as it pertains to a monodisperse system in the approximation of free transport of vibrationally excited molecules to the surface of aerosol particles and considers the existence of vibrational and molecular exchange in a heterogeneous mixture. Based on the analysis of quasi-stationary heterogeneous relaxation regimes, an estimate is given for the catalytic action of aerosol particles on the relaxation processes in a gas. The potential for using aerosol particles as functional elements of a GDL gain medium is discussed.

2.1. *Mechanism for Surface Deactivation of Gas Molecules*

It has been experimentally established that the collision of gas molecules with a surface is highly effective in regard to vibrational relaxation processes. For homonuclear molecules, collisions with a surface are usually characterized by the vibrational transition probabilities $P^s \sim 10^{-4}$–10^{-2}, and for dipole diatomic and polyatomic molecules $P^s \sim 10^{-2}$–1. There are few basic mechanisms of surface deactivation for gas molecules. They are usually divided by type of interaction as collision and adsorption mechanisms [61, 70]. The first mechanism includes *VT*- and *VV'*-processes leading to vibrational deactivation of gas molecules at the moment they collide with the surface, and the second mechanism leads to deactivation in the adsorption stage, which occurs as a result of the molecule adhering to the surface.

From vibrational relaxation theory it is known [71] that, for the process of collisional molecular deactivation, only strong collisions are important which can be characterized by the energy

$$E_* = \chi T \gg T, \qquad \chi = \left(\frac{\pi \mu a^2 \Delta E_v^2}{2\hbar^2 k T}\right)^{1/3},$$

where ΔE_v is the energy liberated when the molecule is deactivated, μ is the reduced mass of the colliding molecules, and a is the effective radius of the repulsive forces. The fraction of such collisions in a gas is small, therefore, the probability of vibrational deactivation of the molecules, averaged over energy, turns out to be low:

$$\bar{P} = \langle P(E) \rangle \sim \exp(-3\chi). \tag{25}$$

The existence of long-range attractive forces between molecules accelerates their motion during the collision, which leads to an increase in the vibrational transition probability. If the attraction energy D for collisions in a gas is not large (D_0 ~ 100-300 K), since it is determined by pair-wise molecular interaction, then the attraction energy can be significant for molecules colliding with a surface (D_s ~ 1000-2000 K), as in this case the intermolecular interaction behaves in a collective manner at large distances.

In order to account for the effect of attraction when calculating the collisional deactivation probability for gas molecules at a surface [68, 70], we shall make use of the fact that a vibrational transition takes place primarily in the region where the repulsive force operates, where the interaction of the incident molecules with surface atoms or molecules is pair-wise. In this case, the role of attraction reduces to imparting additional kinetic energy to the gas molecule that is equal in magnitude to the potential depth: $E' = E + D_s$. Then, representing the interaction potential in the corresponding manner [72]:

$$V_s = V_0 \exp(-x/a) - D_s, \tag{26}$$

for the collisional deactivation probability of the molecules we have

$$P^{cs} = \exp\left(\frac{D_s}{T}\right) \int_{D_s}^{\infty} P(E) \exp\left(-\frac{E}{T}\right) \frac{dE}{T}, \tag{27}$$

where $P(E)$ is the transition probability for a repulsive potential [73]. For fairly small potential well depths ($D_s < E_*$), this expression reduces approximately to

$$P^{cs} = \bar{P} \exp(D_s/T). \tag{28}$$

From here it follows that at low temperatures, the collisional molecular interaction at the surface is characterized by the vibrational deactivation probabilities, which can significantly exceed the probability of exchange relaxation P^0 of gas molecules.

If the collisional interaction process for a vibrationally excited molecule with a surface is not accompanied by deactivation, then the molecule can lose its excitation as a result of adsorption. The adsorption deactivation process may be associated with both the direct adherence of an excited gas molecule and with the resonant transfer of its quantum of energy to adsorbed molecules. In both cases, the vibrational-translational relaxation of molecules at the surface is the final stage of adsorption deactivation of the gas molecules. For a small degree of surface coverage by the adsorbed component, it is possible to neglect the exchange channel of adsorption deactivation of molecules. In this case, the probability of adsorption deactivation of gas molecules P^{as} as calculated for a single collision with a surface is determined in the following manner [61, 70]:

$$P^{as} = \zeta[1 - \exp(-\tau_a Q)], \tag{29}$$

where ζ, τ_a are the sticking probability and the adsorption time of excited gas molecules on the surface, respectively, and q is the extinction rate of the intrinsic vibrations of the adsorbed molecules. Since the lifetime of the molecules in the adsorbed state [74]

$$\tau_a = \nu^{-1} \exp(D_s/T)$$

at low temperatures greatly exceeds the period of the intermolecular vibrations ν^{-1}, even a small decrease in the extinction of the intrinsic vibrations of molecules on the surface Q/ν can yield a high probability of adsorption deactivation of the gas molecules. Using the rate $\nu\bar{P}$ as an estimate of the value of Q, at which vibrational-translational relaxation of molecules takes place during binary collisions in a liquid, it results that in the case $Q\tau_a \ll 1$ the adsorption component of the probability of surface deactivation can be comparable with the collisional component:

$$P^{as} \simeq \zeta\bar{P} \exp(D_s/T). \tag{30}$$

Note that for nonuniform surfaces there are different energy parameters of the molecules' attractive potential characterizing the processes of collisional and adsorption interaction. During adsorption, the molecules tend to choose the deepest location where the attractive potential is high, whereas in a collision with the surface this choice is random. Therefore, it is wise to assume that at low temperatures, when the sticking probability should be fairly high ($\zeta \sim 0.1-1$), vibrational deactivation of gas molecules on a nonuniform surface is determined by the adsorption interaction, and on a homogeneous surface it is determined by collisional interaction.

Above we did not consider the possibility of the deactivation of gas molecules A_1 as a result of the transfer of vibrational excitation by molecules of the condensed phase A_L on the surface. Vibrational exchange between these molecules can occur during the brief collision stage of the interaction and during prolonged adsorption. For reasons already known, the probability of this process at the surface P^{s}_{1L} at low temperatures usually significantly exceeds the same probability P^{0}_{1L} in the gaseous phase. An exception, perhaps, is the exchange by vibrational quanta between molecules A_1 and A_L, whose intrinsic vibrational frequencies are close: $|\omega_1 - \omega_L| < \bar{v}/a$. In this case, as is known from Ref. [71], the force is not important, but the duration of the interaction of the colliding molecules is. Therefore, the probability of resonant vibrational exchange P^s_{1L} at the surface can be significantly greater than the probability corresponding to collisions in a gas P^0_{1L}, due only to the adsorption component which does not exceed ζ.

2.2. Kinetics of Vibrational Relaxation of Molecules in a Gas-Aerosol System

We shall examine the fundamental features of the kinetics of molecular vibrational relaxation in heterogeneous systems. We shall do the analysis as an illustration of a biphase system of the gas-aerosol type, which is a mixture of two types of monomers (A_1, A_2) identical in terms of the number of complexes A_L consisting of molecules of the same sort as A_2. The composition and other determining parameters of the heterogeneous mixture shall be assumed to be constant. The interaction of the complexes in the heterogenous relaxation process for molecules of the gas and condensed phases shall be neglected. This is possible if the collision frequency of the complexes between themselves Z_{LL} in a biphase system is much less than their collisions Z_{Lm} with both types of monomers ($m = 1, 2$):

$$Z_{LL}/Z_{Lm} \simeq 4\sqrt{2}\, m_L^{-3/2}\, \kappa_L/\kappa_m \ll 1,$$

where m_L is the number of molecules in the complex and κ_m is the relative concentration of molecules of the mth component in the mixture. The interaction of monomers with the complexes shall be examined in the assumption that the size of the particles and the mean distance between them are small in comparison with the mean-free-path length of the monomers in the gas.

In these approximations, the description of vibrational relaxation of molecules in a gas-aerosol system turns out to be similar in many ways to that which is usually used in the case of homogeneous molecular mixtures. The main difference consists only in the fact that, in heterogeneous mixtures, the exchange of vibrational energy between different components can be caused not only by the transfer of quanta (*VV'*-process), but also the exchange of molecules (*MM'*-process). The potential for molecular exchange is related to the dynamic character of the phase equilibrium that is established between components, which differ by aggregate state, but are identical in molecular type. Such components shall be called coupled components. In the heterogeneous mixture being examined, the gaseous ($m = 2$) and condensed ($m = L$) components, consisting of molecules of variety A_2, are coupled, as are the gas ($m = 1$) and adsorbed ($m = 0$) components, consisting of molecules of variety A_1.

In describing vibrational relaxation, we shall start by assuming that the vibrational levels of the diatomic molecules are in accordance with a Boltzmann distribution. Then the terms that are related to *VT*-, *VV'*-, and *MM'*-processes in the balance equations for the number of quanta in a calculation for one molecule of any component

$$de_m/dt = (de_m/dt)_{VT} + (de_m/dt)_{VV'} + (de_m/dt)_{MM'} \tag{31}$$

can be represented in the form [75]

$$\begin{aligned} &(de_m/dt)_{VT} = -R_m(e_m - \bar{e}_m), \\ &(de_m/dt)_{VV'} = -\sum_{n \neq m} R_{mn}[e_m(e_n + 1) - \varphi_{mn} e_n (e_m + 1)], \\ &(de_m/dt)_{MM'} = -r_{mn}(e_m - e_n), \end{aligned} \tag{32}$$

where

$$e_m = [\exp(\omega_m/T_m) - 1]^{-1}; \qquad \varphi_{mn} = \exp[-(\omega_m - \omega_n)/T];$$

ω_m, T_m are the frequency (natural) of the vibrations and the vibrational temperature of the molecules of the mth component, and $\bar{e}_m$ is the value of e_m at $T_m = T$.

The expressions for the rates of *VT*- and *VV'*-processes can best be represented in the form which allows gas-kinetic treatment of the interaction of molecules in a heterogeneous mixture:

$$R_m = Q_m + \sum_{n \neq m} P_m^{(n)} \nu_{mn}, \qquad R_{mn} = P_{mn}^{(n)} \nu_{mn}, \tag{33}$$

where Q_m is the rate of vibrational-translational relaxation of the mth component molecules when they interact with each other; $P_m^{(n)}$, $P_{mn}^{(n)}$, or concisely, $P_{m,n}^{(n)}$, is the average probability of *VT*- or *VV'*-deactivation of molecules of the mth component when colliding with molecules of the nth component; and ν_{mn} is the frequency of the corresponding collisions. The value of this notation consists in the fact that it permits finding the relationship between the rates of *VV'*-processes directly from the principle of detailed equilibrium for homogeneous mixtures of similar composition:

$$P_{mn}^{(n)} = \varphi_{mn} P_{nm}^{(m)}, \qquad \nu_{mn}/\nu_{nm} = \kappa_n/\kappa_m. \tag{34}$$

However, it is necessary to keep in mind that in a heterogeneous mixture, the interaction between component molecules found in different aggregate states does not take place uniformly over the entire volume, but is localized at the surface and, consequently, is not possible for some of the molecules. Therefore, the quantities $P_{m,n}^{(n)}$ and ν_{mn} in general can only effectively characterize an elementary process, determining its rate as an average for molecules of a given component.

As an example, we shall determine these values for the interaction of condensed and adsorbed component molecules with gas phase molecules. Let the degree of surface coverage of the complexes by adsorbed molecules be small, i.e.,

$$\theta \simeq \frac{\kappa_0 m_L}{\kappa_L m_L^{(s)}} \ll 1,$$

where $m_L^{(s)}$ is the number of molecules making up the surface of the complex. Then the probability $P_{m,n}^{cs}$, calculated in correspondence with the collisional mechanism of VT- or VV'-deactivation, can be taken for $P_{m,n}^{(n)}$. Since

$$\nu_{1L} = Z_{1L}, \quad \nu_{10} = \theta Z_{1L}, \tag{35}$$

then according to Eq. (34)

$$\nu_{L1} = Z_{L1}/m_L, \quad \nu_{01} = Z_{L1}/m_L^{(s)}. \tag{35}$$

Actually taken into account here is the fact that, although when monomers collide with a complex only one of its molecules is directly subject to collisional deactivation, they can all participate indirectly in this process. This is due to the presence of resonant vibrational exchange that establishes an equilibrium distribution of quanta inside the complexes.

Determining the probability and frequency characteristics of the interaction taking place on the surface of the complexes between molecules and the adsorbed and condensed components is somewhat more complex in the framework of a gas-kinetic treatment. If we assume, however, that a migration mechanism for the extinction of the natural vibrations of adsorbed molecules is responsible for vibrational relaxation [70], then in this case it is possible to confine ourselves to the binary collision approximation. Then, assuming the frequency of the corresponding collisions $\nu_0 L$ is equal to the frequency of the jumps

$$\nu_h = \nu \exp(-D_h/T)$$

and assuming that the motion of the adsorbed molecules as they migrate along the surface is classical, we get

$$P_{0,L}^{(L)} = P_{0,L}^{hs} = \exp\left(\frac{D_h}{T}\right) \int_{D_h}^{D_s} P_{0,L}(E) \exp\left(-\frac{E}{T}\right) \frac{dE}{T}, \tag{37}$$

where D_h is the height of the activation barrier separating the stable positions of the adsorbed molecules on the surface of the complex. Allowed for in this expression is the fact that as a molecule with energy $E > D_s$ moves, it leaves the surface and thereby is excluded from the migration deactivation process.

As for the molecular exchange rates r_{mn}, in the case of coupled gaseous and adsorbed components, they are determined by the following relations:

$$r_{10} = \zeta_1 Z_{1L}, \quad r_{01} = (\tau_a)^{-1}. \tag{38}$$

Assigned in the same way are the molecular exchange rates for the coupled gaseous and condensed components. However, ζ_1 and τ_a should be understood as quantities characterizing the processes of molecular condensation and vaporization, respectively. It is obvious that for uncoupled components $r_{mn} = 0$.

Different regimes of heterogeneous relaxation are possible depending on the relationship between the rates of *VT*-, *VV'*-, and *MM'*-processes for the gaseous, adsorbed, and condensed components in a gas-aerosol system. The most typical are those realized when the rate of the indicated processes turns out to be significantly higher for the adsorbed component than for the others. This is due to the relatively small concentration of adsorbed molecules in the mixture ($\kappa_0 << \kappa_n$, n = 1, 2, L), and also to the high efficiency of the collisional and migrational mechanisms of surface *VT*- and *VV'*-relaxation.

We shall show that under such conditions, heterogeneous relaxation of molecules in the gas and condensed phases proceeds for a quasi-steady distribution of vibrational quantum with the adsorbed component. We shall write Eq. (31) in linearized form:

$$de_m/dt = -\tilde{R}_m (e_m - \tilde{e}_m), \tag{39}$$

where

$$\tilde{e}_m = (1/\tilde{R}_m) R_m \bar{e}_m + \sum_{n \neq m} (R_{mn} \varphi_{mn} e_n + r_{mn} e_n)];$$

$$\tilde{R}_m = R_m + \sum_{n \neq m} R'_{mn}; \quad R'_{mn} = R_{mn} + r_{mn}.$$

Here the parameter $\tilde{e}_m$ acts as the relatively equilibrium value for e_m, which characterizes the vibrational state of molecules of the mth component that is established for

$$\frac{1}{\tilde{e}_m}\left|\frac{d\tilde{e}_m}{dt}\right| \ll \tilde{R}_m. \tag{40}$$

Since the change in $\tilde{e}_m$ occurs at a rate that does not exceed the value

$$\frac{1}{\tilde{e}_m}\left|\frac{d\tilde{e}_m}{dt}\right| \simeq \sum_{n\neq m} \tilde{R}_n,$$

then Eq. (40) is possible if $\tilde{R}_m >> \tilde{R}_n$, which also needs to be shown.

The existence of a quasi-steady regime of vibrational relaxation for one of the components, in this case the adsorbed component, permits us to significantly simplify the analysis of the system of Eqs. (31) and (32). Formally, this is done by excluding the equation for the adsorbed component from the system and replacing the value e_0 by its quasi-stationary value $\tilde{e}_0$ in the right-hand-side of the remaining equations. Performing this procedure and neglecting the difference in the natural vibration frequencies of the molecules of the coupled components, we obtain in linear approximation

$$de_m/dt = -K_m(e_m - \bar{e}_m) - \sum_{n\neq m} K_{mn}(e_m - \varphi_{mn}e_n), \tag{41}$$

where

$$K_m = R_m + R'_{m0}R_0/\tilde{R}_0, \quad K_{mn} = R'_{mn} + R'_{m0}R'_{0n}/\tilde{R}_0. \tag{42}$$

In the expressions for the heterogeneous *VT*- and *VV'*-relaxation rates of gaseous and condensed phase molecules, the first terms correspond to *VT*- and *VV'*-processes caused by indirect molecular interaction of the components among themselves, and the second terms correspond to relaxation processes taking place with the participation of the adsorbed molecules.

We shall discuss the most qualitatively interesting consequence of the obtained results pertaining to the role of the adsorbed component in surface relaxation of gas molecules.

Section 2.1 determined the probability of adsorption deactivation of gas molecules. It was assumed there that during adsorption the molecule being deactivated is not able to transfer its excitation to gas phase molecules, i.e., $\tau_a R_{01} << 1$. In practice, this is possible at low pressures or relatively high temperatures. Separating adsorbed molecules into an independent component of a heterogeneous mixture with corresponding inclusion in the examination of molecular and vibrational exchange permits removing this constraint on the description of the adsorption mechanisms of the deactivation of gas molecules. One can be convinced of this by an example of kinetic relation for $P^{as}_{1,L}$, following from Eq. (42):

$$P^{as}_{1,L} = R'_{10} R_{0,L} / Z_{1L} \tilde{R}_0. \tag{43}$$

Here we allowed for the fact that for a phenomenological determination of the heterogeneous relaxation rates

$$K_1 = Q_1 + P^s_1 Z_{1L}, \quad K_{1L} = P^s_{1L} Z_{1L} \tag{44}$$

the probability of surface *VT*- or *VV'*-deactivation of gas molecules is $P^s_{1,L} = P^{as}_{1,L} + P^{cs}_{1,L}$.

We shall expand on Eq. (43) in the case of a single-component gas, when the adsorption deactivation process for A_1 molecules can occur either along the *VT*-channel, or due to vibrational exchange between adsorbed molecules and condensate molecules. Then, neglecting the corresponding terms in $\tilde{R}_0$ (R_{0L} for the *VT*-process and R_0 for the *VV'*-process) and recalling that in the state of phase equilibrium between coupled gas and adsorbed components

$$R_{10}/R_{01} = \kappa_0/\kappa_1 = r_{10}/r_{01},$$

we have

$$P^{as}_{1,L} = \zeta_1 \tau_a R_{0,L} \left(1 + \frac{\tau_a R_{0,L}}{1 + \tau_a R_{01}}\right)^{-1}. \tag{45}$$

From here it follows that the kinetic definition (43) transforms into Eq. (29) only for $\tau_a R_{01} \ll 1$ (low pressures). In the opposite case, it usually does not correspond to Eq. (29) and describes the so-called exchange mechanism of adsorption deactivation of gas molecules [55, 61], which is realized at high pressures or low temperatures.

However, it turns out that if in Eq. (29) we understand ζ to be the quantity θP^{cs}_{10} and τ_a to be $(R_{01})^{-1}$, respectively, then it can also correctly describe the given case. Actually, this means that the parameters ζ and τ_a in Eq. (29) are now characteristics of the process of adsorption of quanta, and not of molecules. The adsorption mechanism is treated as a two-stage process, for which vibrational deactivation of gas molecules emerges as a result of the capture and escape of quanta during adsorption on the surface. Analysis shows that such a treatment also turns out to be valid in general, when the transport of vibrational excitation of gas molecules to the surface can be caused both due to the transfer of quanta by adsorbed molecules and by the adherence of excited molecules. Then, for the sticking probability of a quantum in relations of the type of Eq. (29), we should use the quantity

$$\hat{\zeta} = \zeta_1 + \theta P_{10}^{cs}, \tag{46}$$

and for the adsorption time we should use

$$\hat{\tau}_a = \tau_a/(1 + \tau_a R_{01}). \tag{47}$$

We shall now pause on a few features of the quasi-steady regime of heterogenous molecular relaxation in disperse systems with a small admixture of condensate ($\kappa_L << \kappa_1, \kappa_2$). In this case, heterogeneous relaxation of gas phase molecules A_1 and A_2 should proceed with a quasi-steady distribution of quanta encompassing not only the adsorbed component, but the condensed component of the mixture as well. Then

$$e_L = \tilde{e}_L = \frac{1}{\tilde{K}_L}(K_L \bar{e}_L + \sum_{n \neq L} K_{Ln} \varphi_{Ln} e_n). \tag{48}$$

where $\tilde{K}_L = K_L + \sum_{n \neq L} K_{Ln}$. Using Eq. (48), system of Eqs. (41), describing the kinetics of heterogeneous relaxation of gas phase molecules acquires the form

$$de_m/dt = -\Gamma_m (e_m - \bar{e}_m) - \Gamma_{mn} (e_m - \varphi_{mn} e_n), \tag{49}$$

where

$$\Gamma_m = K_m + K_{mL} K_L/\tilde{K}_L; \qquad \Gamma_{mn} = K_{mn} + K_{mL} K_{Ln}/\tilde{K}_L. \tag{50}$$

In the relations derived for Γ_{mn}, the first terms correspond to heterogeneous *VV'*- or *VT*-relaxation proceeding in a direct channel, and the second terms correspond to exchange through the condensate molecules.

We shall clarify how effectively both channels operate for the catalytic action of aerosol particles on vibrational-translational relaxation of gas molecules. Let the condensate particles consists of the same sort of molecules as the gas and the liquid drops have identical volume $m_L b$, where $b = (2/3)\pi d^3$ is the intrinsic volume of a molecule in the liquid, and d is the molecule's gas-kinetic diameter. Then the expression corresponding to the component Γ_1, corresponding to the direct channel of heterogeneous *VT*-relaxation ($\Gamma_1^{VT} = K_1$), can be given in the following form:

$$\Gamma_1^{VT} = n(\kappa K_1^{(0)} + \kappa_L K_1^{(s)}), \tag{51}$$

where

$$K_1^{(s)} = P_1^s Z_{1L}/\kappa_L n = \pi d^2 \bar{v} P_1^s (1 + \sqrt[3]{4m_L})^2/4\sqrt{2}m_L;$$

$K_1^{(s)}$, $K_1^{(0)}$ are the surface and bulk VT-deactivation gas constants for gas molecules, respectively, n is the total concentration of molecules in the mixture, and $\bar{v}$ is the thermal velocity of the molecules.

The exchange component Γ_1 can be represented similarly:

$$\Gamma_1^{VV'T} = n\kappa_L K_{1L}^{(s)} \beta_L. \tag{52}$$

The quantity $\beta_L = K_L/\tilde{K}_L$ represents the probability of a quantum escaping during its one-time stay in a condensate particle. As a function of the size of the complex for $m_L \gg 1$, it can be determined as follows:

$$\beta_L = (1 + \sqrt[3]{m_*/m_L})^{-1}, \quad m_* = (\nu_{11} P_{L1}^s/2Q_L)^3. \tag{53}$$

It is assumed that the rate of vibrational-translational relaxation of molecules in liquid phase particles is independent of their size m_L and equal to Q_L. In this approximation

$$\Gamma_1^{VV'T} = \kappa_L Q_L (1 + \sqrt[3]{m_L/m_*})^{-1}. \tag{54}$$

The form of the dependence of $\Gamma_1^{VV'T}$ on m_L is caused by the competition of two factors. On one hand, with an increase in m_L the intensity of vibrational exchange between gaseous and condensed components decreases, and on the other, the probability of losing quanta to condensate particles increases. As long as the size of the particles is small in comparison with m_*, both factors completely compensate for each other, and therefore the rate of heterogeneous molecular relaxation in the exchange channel turns out to be independent of the dispersion of the biphase system. On making the transition to the $m_L > m_*$ region, where exchange deactivation of gas molecules in complexes takes on, as in the case of macroscopic particles, a irreversible character ($\beta_L = 1$), the efficiency of this relaxation channel begins to drop in agreement with how the exchange rate constant $K_{1L}^{(s)}$ depends on m_L.

We shall now perform a comparative calculation of the efficiency of both channels of heterogeneous VT-relaxation of gas molecules. Since direct VT-deactivation of gas molecules in particles of intrinsic condensate can be caused only by the collisional mechanism, then according to Eq. (27)

$$P_1^s/P_1^0 \simeq \exp[(D_s - D_0)/T].$$

Assuming $D_s = 4D_0$, it results that for CO_2 molecules, for example, at $T = 100$ K $P_1^s/P_1^0 = 3 \cdot 10^2$. For this difference in the probabilities of surface and bulk VT-deactivation of gas molecules, the ratio of the corresponding relaxation rates for $\kappa_L = 0.1$ and $m_L = 10^2$ is $\Gamma_1^{VT}/\Gamma_1(\kappa_L{=}0) = 5$. If the exchange channel of surface deactivation is included, then the catalytic action of condensate particles on the vibrational relaxation of gas molecules can appear more distinctly. If we assume that vibrational relaxation of molecules in condensate particles occurs at the rate Q_L, corresponding to the mechanism of binary collisions in a liquid [76, 77], then for the bending vibrations of CO_2 for $n = 10^{18}$ cm^{-3} ($m_* \simeq 10^3$) and for the aerosol parameters indicated above, the value of $\Gamma_1^{VV'T}/\Gamma_1(\kappa_L{=}0)$ is 10^3.

2.3 Using Aerosol Particles in a Gas-Dynamic Laser Gain Medium

The estimate of the effect of aerosol particles on the kinetics of vibrational-translational relaxation molecules given above shows that they can be a source of significant relaxation energy losses in GDL gain media. But the catalytic activity of the aerosol particles is not manifested only in this. A no less significant, but an already positive role for aerosol particles is as the catalyst of nonresonant vibrational exchange between gas molecules. This is especially important when the direct exchange of quanta between A_2 molecules, carrying the primary reserve of vibrational energy of the gain medium (energy carrier) and amplifying the radiation by A_1 molecules due to the relatively large resonance defect ($\Delta\omega_{21} > T$) has a low probability and, therefore, is not able to provide a high rate of nonresonant pumping of the A_1 molecules. Acting as an intermediary for the quantum exchange between A_1 and A_2 molecules, aerosol particles can greatly shorten the characteristic time τ_{12} of transferring vibrational energy from A_2 molecules to the radiating component of the gas and thereby facilitate more effective use of the advantages of nonresonant pumping at low gas temperatures.

We shall examine this effect in more detail under conditions when the narrowest place in the vibrational exchange along the A_1-(A_L)-A_2 channel is the process for transferring quanta between A_1 and A_L molecules. This case is most typical when using aerosol particles consisting of the same sort as the gain medium energy carrier A_2 molecules.

Let the external parameters (mass, diameter) of the mixture's molecules be identical. Then, assuming that the fraction of condensate of component A_2 is small ($q = \kappa_L/\kappa_2 << 1$), according to Eq. (50) we get

$$\frac{\Gamma_{12}(q)}{\Gamma_{12}(0)} \simeq 1 - q + \frac{q}{2\sqrt[3]{m_L}} \frac{P^s_{1L}}{P^0_{12}}(1-\beta_L). \tag{55}$$

From here it is clear that the smaller the condensate particle size is, the more strongly the catalytic effect should appear. This is caused not only by the increase in the rate of vibrational exchange between gas and condensed phase molecules, but also by the increase in the probability of losing quanta as they are transferred along the heterogeneous channel. If $m_L << m_*$, then the loss of quanta due to relaxation losses in condensate particles can be neglected. In this case, the catalytic action of the aerosol on the vibrational exchange in a gas turns out to be roughly the same as on direct VT-relaxation of A_1 molecules.

We shall make a quantitative estimate of the catalytic action of the aerosol on the vibrational exchange in a $CO_2(1)$-$O_2(2)$ mixture, in which the O_2 molecules are the vibrational energy carrier for the coupled modes of CO_2. This example is interesting in relation to the possibility of making a gas-dynamic CO_2-laser operating at the 03^10-10^00 transition. Assuming that the adsorption mechanism of VV'-deactivation is responsible for vibrational exchange between CO_2 molecules and O_2 condensate particles ($\omega_2 - \omega_1 = 270$ K), and estimating R_{0L} in Eq. (45) in the approximation of binary molecular collisions in a liquid with $\nu_{0L} = \nu$, it results that if $\zeta_1 = 1$, $D_s = 6D_0$ ($D_0 = \sqrt{D_0(1)D_0(2)} = 155$ K), then the value of $P^s_{1L}/P^0_{12} = 3{\cdot}10^3$ at $T = 100$ K ($P^0_{12} = 10^{-5}$ [52]), and $\Gamma_{12}/\Gamma_{12}(0) = 10$ for $q = 0.03$, $m_L = 100$. It is significant that this acceleration of vibrational exchange is possible on a background of a narrow level of relaxation losses of quanta in condensate particles ($\beta_L <<1$). This is because of the small extinction rate of the natural vibrations of O_2 molecules in the liquid phase ($Q \sim 10^2$ s^{-1} [78]), or more accurately, with the large critical size of the condensate particles ($m_* >> m_L$).

Even more promising from the point of view of exploiting the advantages of nonresonant pumping of the radiating component molecules are gain media in which condensate particles act as the radiator. We shall assume that the carrier of vibrational energy in this type GDL are molecules of the slowly relaxing gaseous component, and emission of radiation is achieved at the vibrational transitions of the natural vibrations of condensate phase molecules. In this case, the transfer of vibrational energy to molecules of the radiating component at low temperatures can occur two to three orders of magnitude faster than in the analogous gaseous mixture. However, due to the large radiation line width ($\overline{\Delta\nu} \approx 5$–$10$ cm^{-1}), corresponding to vibrational bands of condensate phase molecules, and the comparatively low concentration of condensate molecules in the mixture, the gain coefficient in such a biphase gain medium should be small: $\alpha \sim 10^{-4}$ cm^{-1}, if $n_L \sim 10^{17}$ cm^{-3}, $e_L \sim 1$, $\overline{\Delta\nu} \sim$ 10 cm^{-1}, $A_{21} \sim 10$ s^{-1}, and $\lambda \sim 10$ μm.

In connection with this, we shall discuss in more detail the GDL scheme based on a biphase medium, in which aerosol particles are the energy carrier and the radiating component gas molecules are the gain medium. It is suggested that pumping of the gas molecules is accomplished in the process of vibrational exchange with the aerosol molecules, forming the basis of the biphase system ($\kappa_L >> \kappa_1, \kappa_2$). The particle size should be large in order for their coagulation process to proceed slowly ($Z_{LL} \sim \pi d^2 \bar{v} n_L m_L^{-5/6}$), and also in order for the heterogeneous processes, caused by surface VT-relaxation of the aerosol energy carrier molecules, to be insignificant. The latter remark mainly applies to solid phase particles, for which the bulk component of the extinction rate of the natural molecular vibrations can be neglected in comparison with the surface component. This is because vibrational molecular relaxation due to multiphase transitions occurs extremely slowly [77].

The upper limit on the particle size is imposed by the need to provide a high rate of nonresonant pumping of the nonradiating component molecules of the gas, and also by the condition $m_L << m_*$. This latter limitation is primarily important for a liquid energy carrier and is necessary for the energy losses in bulk relaxation of aerosol molecules to be significantly less than the (useful) losses in pumping gas molecules. Starting from the aforementioned acceptable ratio for a CO_2:O_2 = 1:100 mixture, for example, the following parameters of the aerosol energy carrier can be assumed for O_2: $n_L \sim 10^{21}$, $m_L \sim 10^9$. In this situation, the relative losses are $\beta_L \sim 0.1$ (corresponding to $m_* \sim 10^{12}$ and a pumping rate of $J_1 \sim 3 \cdot 10^6$ s^{-1}), and a coagulation length in a supersonic flow of $L \sim 50$ cm.

A significant advantage of this laser system in comparison with that examined above is not only the narrow gain bandwidth $\overline{\Delta\nu} \sim 10^{-2}$-$10^{-1}$ cm^{-1}, typical for the usual homogeneous GDL, but the significantly larger energy capacity of the gain medium as well. For the aforementioned parameters of an aerosol energy carrier based on O_2, the energy density in a biphase gain medium for $e_L = 0.3$ ($\omega_L = 2230$ K) is 10 J/cm^2, which is two to three orders of magnitude greater than in homogeneous GDL. In this sense, GDL with an aerosol energy carrier are favorably distinguished from gas lasers in terms of their energy characteristics and from lasers based on condensed media in terms of their spectral characteristics. In addition, the use of a condensed material to accumulate the vibrational energy permits one to significantly broaden the potentials of different combined (electrical, optical) methods of pumping the molecules in GDL gain media.

We shall estimate the possible parameters of a gas-dynamic CO_2-laser with an aerosol energy carrier based on liquid nitrogen with electron beam pumping. Let the energy of the electrons be $E = 0.5$ MeV and the current density of the beam be $I = 100$ A/cm^2. Then the depth h that the electron beam penetrates into the biphase flow with an aerosol mass density of ρ_L (g/cm^3), corresponding to $n_L = 10^{20}$ cm^{-3}, according to Ref. [79] equals $h = (0.55E - 0.16)/\rho_L = 25$ cm. The rate of pumping the nitrogen energy carrier molecules due to different recombination processes in the condensed phase [77] for an efficiency of coupling energy into the vibrational degrees of freedom of $\delta = 0.2$ is $J_L \simeq \delta EI/hn_L \simeq 2.5 \cdot 10^4$ eV/s. For this pumping rate, the level of vibrational excitation of N_2 molecules during the flight of particles through the interaction zone with a beam 1 cm long for a flow rate of $2 \cdot 10^5$ cm/s is $e_L = J_L t/\omega_L \simeq 0.42$. The nonequilibrium supply of vibrational energy of 480 J/g corresponding to this value can be stored for a long time ($Q_L \sim 1$-10 s^{-1}) by liquid nitrogen molecules. This fact permits using an aerosol energy carrier based on nitrogen in a gas-dynamic CO_2-laser, operating on the principle of biphase flow mixing [65, 80]. If the mixing of the N_2 aerosol particles is accomplished with a flow of a gaseous mixture of CO_2:He = 1:4 at a pressure of 0.1 atm and a temperature of 100 K, the density of the population inversion at the CO_2 00^01-10^00 transition is $4 \cdot 10^{17}$ cm^{-3}, and the gain coefficient at a wavelength of 10.6 μm is $\alpha \simeq 4 \cdot 10^{-2}$ cm^{-1}.

3. Heterogeneous Processes in Gas-Dynamic CO_2-Lasers

Any gas, if it is cooled adiabatically, can go into a supersaturated state and begin to condense. The process of condensing a supersaturated gas is a result of spontaneous formation and growth of small-sized particles nucleating from the liquid phase. In the gain medium of a CO_2-GDL, saturation conditions for water vapor are usually reached when cooling the gaseous mixture down to a temperature of $T_\infty \approx 150$ K. The appearance of condensate particles at low temperatures can lead to significant acceleration of vibrational relaxation of the molecules and, consequently, to additional losses of vibrational energy stored in the GDL gain medium. In connection with this, the task arises to concurrently examine the kinetics of the relaxation and condensation processes in the supersonic flow of gaseous mixtures that are used in CO_2-GDL.

The question of the effect of water vapor condensation on the operation of a gas-dynamic CO_2-laser was first raised in Ref. [81], where the authors managed to experimentally observe a sharp drop in the gain coefficient α at the (00^01)-(10^00) transition in CO_2 in the region of biphase flow of a CO_2-N_2-H_2O-He mixture. The observed effect was attributed to the potential for absorption of λ = 10.6 μm laser radiation by water aerosol particles. In Ref. [68] the suggestion was made that it is caused by the high efficiency of aerosol particles for vibrational deactivation of CO_2 molecules. Later this suggestion received a quantitative foundation in Refs. [82-84]. An estimate of the conditions, under which the optical activity of the aerosol particles becomes significant, was given in Ref. [85].

In this section, the effect of condensation and the heterogeneous losses associated with it on the operation of gas-dynamic CO_2-lasers are studied. Modeling the condensation is conducted in the framework of classical nucleation theory allowing for the most typical features of gas cooling in supersonic flowstreams for GDL. Based on the analysis of different heterogeneous loss channels and comparing the results of modeling heterogeneous relaxation with experimental data, a conclusion is drawn regarding the nature of the activity of water condensate particles on the parameters of the gain media of a typical CO_2-GDL. A similar treatment is performed in relation to the condensation of carbon dioxide gas in a CO_2-GDL operating on the coupled modes of CO_2.

3.1. Modeling Water Vapor Condensation in a Supersonic Jet of Diverging Gas

Gas flow with condensation has already been the subject of intense theoretical and experimental research. However, in view of the complexity of describing the phenomenon of condensation from the position of microscopic theory, an engineering calculation technique is used in the majority of applied papers that is based on the classical drop model of nucleation in a supersaturated vapor [74]. In this approach, it is assumed that the gas-condensate system is in thermal equilibrium, the bulk and surface characteristics of the nucleating centers coincide with the corresponding characteristics for a drop of microscopic size, and the nucleation processes takes place in a gradual scheme of molecular association and behaves in a quasi-stationary manner.

These approximations impose a series of constraints on the conditions of the evolution of condensation in an expanding gas. The most important of these pertains to the gas cooling rate, on which depend the maximum attainable degree of supersaturation in the flow and the minimum size of the nucleation centers that serve as the initial condensation centers. Recall that in classical theory, the nucleation centers are assumed to be only those which are in equilibrium with the supersaturated vapor. According to the Thomson formula, which determines the equilibrium vapor pressure above the surface of an arbitrarily sized drop [74]

$$p_r = p_\infty \exp(2\sigma b/rkT), \tag{56}$$

there corresponds a (critical) radius to this pressure, equal to

$$r_* = 2\sigma b/kT\ln S_\infty, \tag{57}$$

where $S_\infty = p/p_\infty$ is the supersaturation level of the vapor at temperature T, p_∞ is the saturation pressure, σ is the surface tension, and b is the molecule's intrinsic volume in the liquid phase.

For small size particles, a vapor that is supersaturated in the usual sense of the word, i.e., in relation to a macroscopic sized drop, turns out to be unsaturated and therefore the particles evaporate. Conversely, large size particles in an atmosphere that is unsaturated for those particles, therefore, also have a tendency toward systematic growth. Nucleation centers of critical radius form in correspondence with this "narrow range" in the vapor condensation process. The rate of their formation in a supersaturated vapor under stationary conditions is given by the following expression [74]:

$$I_* = c\varsigma\pi r_*^2 \bar{v} n^2 \exp(-{}^4/_3\,\pi r_*^2 \sigma/kT), \tag{58}$$

where

$$c = (8b)^{-1}(kT/\sigma)^{3/2}(\ln S_\infty)^2;$$

ς is the sticking probability, n is the concentration, and $\bar{v}$ is the thermal velocity of the vapor molecules.

For high supersaturation levels ($S_\infty \sim 10^2$-10^3), at which water vapor condensation usually occurs in CO_2-GDL, the critical nucleation center size r_* may be comparable in value to the distance between neighboring molecules of the liquid. In this case, using macroscopic characteristics for a condensation nucleus turns out to be incorrect. In order to eliminate this deficiency, in engineering calculations for the surface tension σ, a correction is introduced to the nucleation center size [86]:

$$\sigma = \frac{\sigma_\infty}{1 + a/r}. \tag{59}$$

The numerical value of the parameter a is chosen such that the free energy of nucleation center formation [74]

$$\Delta\Phi(r) = \frac{4}{3}\,\frac{\pi r^3}{b} kT \ln S_\infty - 4\pi r^2 \sigma, \tag{60}$$

determined using Eq. (59), most accurately corresponds to the calculation of $\Delta\Phi$ in the temperature range of interest. This latter calculation is done on the basis of a specific molecular model of the complex or known experimental data. A calculation of the parameter a with respect to reference data on the equilibrium concentration of water dimers [87] in the 250-300 K temperature range shows that, as a function of the choice of theoretically possible parameters of the dipole-dipole interaction potential of water molecules, it can take on values lying in the interval $a \approx 0.7$-1.5 Å. Similar calculations, obtained from statistical calculation [88] of the concentration of water dimers with a binding energy of 4.8 kcal/mol, give $a = 1.2$ Å.

One more feature of modeling the kinetics of the condensation of water vapor in GDL is related to this high rate of gas cooling. Bear in mind the delay in establishing a quasi-steady nucleation rate that arises with a rapid change of the supersaturation in the flow. In the simplest case, when the transition of the vapor from an unsaturated to a supersaturated state takes place instantaneously, the process of establishing a quasi-steady nucleation rate I_* in the framework of classical condensation theory is described as follows [89]:

$$I(t) \simeq I_* \exp(-\tau_l/t), \tag{61}$$

where

$$\tau_l \simeq (c^2 \zeta \pi r_*^2 \bar{v} n)^{-1}.$$

An estimate of the time lag τ_l under the conditions for water vapor to condense that are typical for CO_2-GDL shows that it can significantly exceed the characteristic time to measure I_*, which is related to the gas cooling:

$$\frac{\tau_l}{I_*}\left|\frac{dI_*}{dt}\right| \gg 1.$$

In connection with this, the calculation of I must be done allowing for nucleation relaxation. An exact solution to this problem on the basis of kinetic analysis of the particle size distribution [90, 91] is extremely laborious, moreover in this case it is hardly valid in view of the absence of reliable information on the processes of growth and vaporization of small-size nucleation centers. Therefore, to model the effect of nucleation relaxation on the condensation kinetics, it is best to confine ourselves to the simplest approximation, determining I from the equation

$$dI/dt = -(I - I_*)/\tau. \tag{62}$$

in which the relaxation time τ is taken to be equal to τ_l.

Figures 10 and 11 show some results of modeling water condensation kinetics in a jet of a gaseous mixture flowing through a slit into a vacuum. The calculations were performed in the relaxation approximation for I according to the usually used [92] scheme of integrating the equations for the kinetics of the formation and growth of the liquid phase in supersonic flowstreams:

$$u(x)\frac{dY_m}{dx} = Y_{m-1}\left(\frac{d\bar{r}}{dt}\right) + 8\pi\frac{r_*^{m-1}}{(m-1)!} IA(x), \tag{63}$$

where $u(x)$ is the speed of motion of the gas in a cross section of the tube of current with area $A(x)$, and $(d\bar{r}/dt)$ is the linear growth rate of a drop with mean radius $\bar{r}$. In the adopted notation, Y_1 determines the concentration

$$N_L = Y_1/8\pi A,$$

Y_3 determines the specific area

$$S_L = \pi\bar{r}^2 N_L = Y_3/4A,$$

Y_4 determines the mass fraction of condensate particles

$$q_L = Y_4\rho_\infty/\rho A,$$

where ρ is the mass density of the vapor at a given cross-section ($q_L = 0$) and ρ_∞ is the density of the liquid phase.

The rate of growth of a drop of supercritical dimensions was calculated in the approximation of thermal equilibrium in a gas-condensate system according to the formula

$$d\bar{r}/dt = \zeta n\bar{v}b[1 - S_\infty^{-1}\exp(2\sigma b/\bar{r}kT)]. \tag{64}$$

The critical size of the nucleation center was found from the condition of the maximum free energy $\Delta\Phi(r)$ for a surface tension that is dependent upon radius r according to the equation

$$r_* = r_*^\infty (1 + \sqrt{1 + 2a/r_*^\infty}) - a, \tag{65}$$

where r_*^∞ is the value of r_* corresponding to $\sigma = \sigma_\infty$. The size of water dimers was assumed to be the minimum for condensation nuclei. The parameter a was assumed to be equal to 1.2 Å, and we used the following temperature dependencies for σ_∞ and p_∞ [90, 93]:

$$\sigma_\infty = (128 - 0{,}19T) \cdot 10^{-7}, \qquad p_\infty = \exp(14{,}46 - 5338/T),$$

where σ_∞ is in J/cm^2 and p_∞ is in atm. The sticking probability for water molecules was taken to be unity.

The gas-dynamic flow parameters were calculated from the central tube with area $A(x)$, corresponding to the isentropic flow of gas from a slit into a vacuum for $\gamma = 1.4$ and $h_0 = 0.04$ cm. It was assumed that, at heating temperatures $T_H \approx 1200$-1800 K that are typical for CO_2-GDL and a relative water vapor concentration of $\kappa \sim 0.01$-0.1 in a CO_2-N_2-H_2O mixture, it is possible to neglect stagnation of the flow during the liberation of heat in the condensation process and the corresponding change in the gas density:

$$\Delta\rho/\rho \simeq \Delta u/u \simeq q_L \kappa L/H << 1,$$

where L is the heat of vaporization and H is the enthalpy of the mixture. This significantly simplifies the calculation, since for the parameters $u(x)$ and $\rho(x)$ that are known for isentropic flows, the temperature and pressure of the mixture in a flow with condensation can be directly found from the energy equation

$$u^2/2 + [(\gamma - 1)/\gamma] RT - q_L \kappa L = \mathcal{H} \tag{66}$$

and state equation

$$p = \rho RT(1 - \kappa q_L). \tag{67}$$

Figure 10 shows typical axial distributions of the parameters characterizing the process of nonsteady condensation of vapors in a freely expanding jet of a gaseous mixture. Also shown for comparison are these same curves calculated in the quasi-steady approximation. As is evident from Fig. 10, for nonsteady condensation of a supersaturated vapor, the buildup in the rate I of the formation of nucleation centers of critical size with increasing supercooling $\Delta T = T_\infty - T$ takes place with some delay caused by nucleation relaxation. In a moving gas, this is expressed as an additional shift of the leading edge of the condensation front relative to the dew point location. The basic shift is determined by the precondensation region in a supersonic flow, where supercooling of the vapor is still too low (ΔT < 50 K) to exhibit the intense ($I \sim 10^{20}$ $cm^{-3} \cdot s^{-1}$) formation of condensation nuclei.

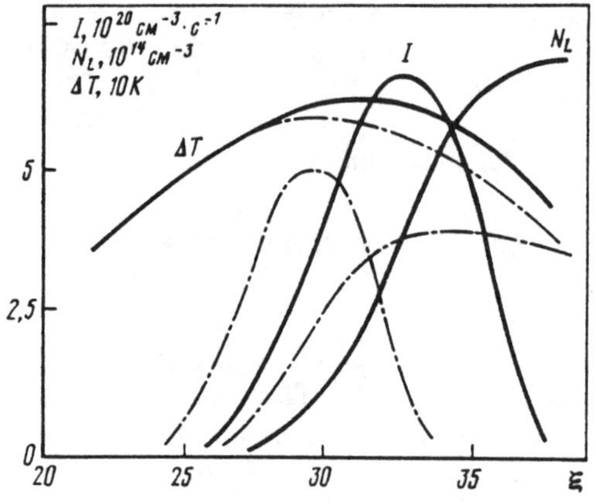

Fig. 10. Axial distributions of the nucleation rate I, particle concentration N_L, depth of supercooling ΔT, specific area S_L, mass fraction q_L, critical radius r_* and mean radius $\bar{r}$ of H_2O condensate particles in a jet for $\kappa = 0.1$, $p_s = 20$ atm, and $T_s = 1300$ K.

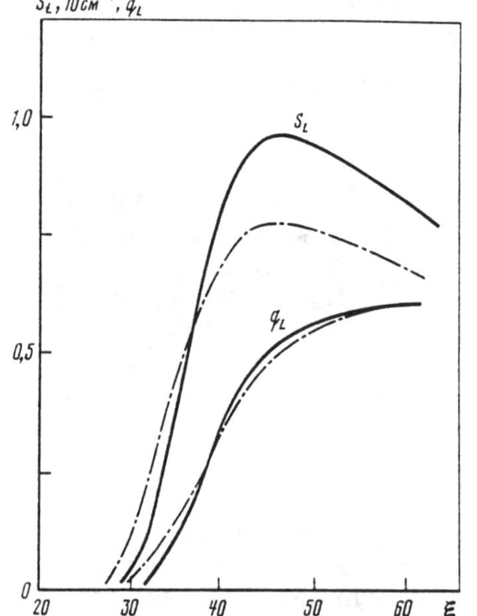

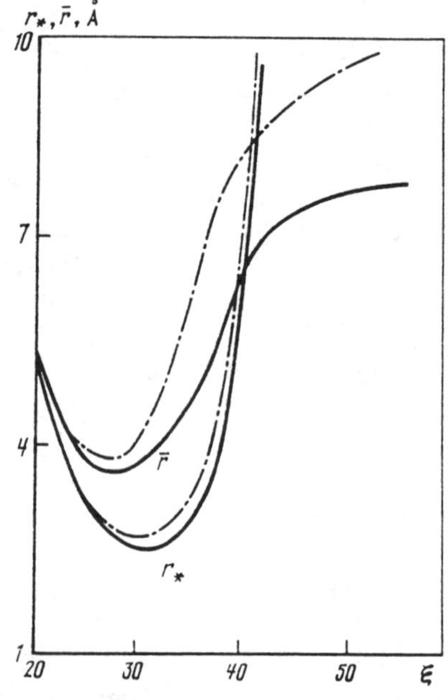

Nucleation relaxation does not solely appears in the additional shift of the condensation front. It also facilitates reaching greater supercooling in an expanding flow in comparison with the quasi-steady condensation regime and, consequently, higher rates of formation and higher concentrations of condensation nuclei. This, as is clear from Fig.

10, in turn leads to increasing the dispersion of the gas-condensate system. It is significant that the mass fraction of condensate here is almost invariant, since the particle size also decreases correspondingly with increasing concentration (see Fig. 10).

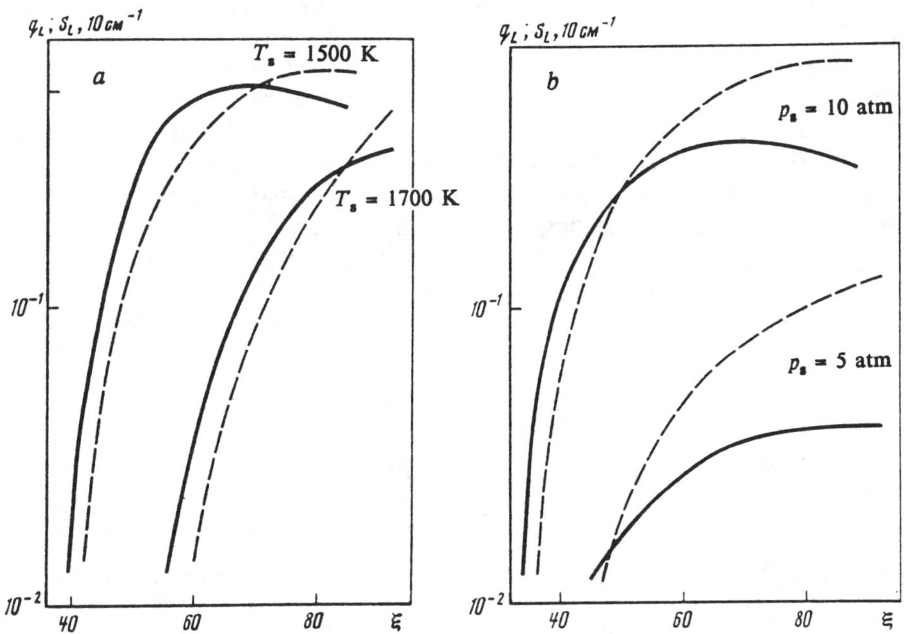

Fig. 11. Axial distributions of the specific area S_L (solid curves) and mass fraction q_L (dashed curves) of H_2O condensate in a jet at different temperatures (*a*) and different pressures (*b*).
a) $\kappa = 0.1$, $n_s = 1.2 \cdot 10^{20}$ cm^{-3}, *b*) $\kappa = 0.1$, $T_s = 1300$ K.

Figure 11 shows the axial distributions of the main gas-condensate system parameters at stagnation temperatures and pressures that are typical for CO_2-GDL. From Fig. 11, *a* it is evident that the condensation front is shifted downstream with increasing stagnation temperature. The concentration N_L, mass fraction q_L, and specific area S_L of the condensate particles decrease. The cause of this is the decrease in the water vapor density $\kappa n_s(T/T_s)^{1/(\gamma-1)}$ in that region of the jet where the gas temperature necessary for intense condensation $T \simeq T_\infty - (\Delta T)_{max}$ is achieved. A decrease in the stagnation pressure or initial water vapor content in the mixture obviously leads to the same result (see Fig. 11, *b*).

3.2. Analysis of the Heterogeneous Loss Mechanisms in a CO_2 Gas-Dynamic Laser

It is known that the attenuation of 10.6 μm laser radiation intensity as it passes through a humid atmosphere is mainly due to the presence of small water aerosol particles in it. This is because water molecules in the condensed phase are much more able to absorb IR-radiation in the $\lambda \simeq$ 8–14 μm region than in the gas phase, where there is a window of transparency to that radiation. In addition, aerosol particles can also cause scattering of the radiation. However, in distinction from a water aerosol existing in an atmosphere, the size of particles formed in the greatest concentration during the homogeneous condensation of water vapor in supersonic GDL flowstreams is usually very small ($\overline{r} \sim 10^{-7}$ cm) in comparison with the CO_2-laser wavelength. As a consequence of this, the fraction of radiant power which is scattered by the aerosol particles [94]

$$Q' = \frac{8}{3}\left(\frac{2\pi\overline{r}}{\lambda}\right)^4 \mathrm{Re}\left(\frac{\hat{m}^2 - 1}{\hat{m}^2 + 2}\right) \tag{68}$$

turns out to be significantly less than the fraction absorbed by them:

$$Q'' = \frac{8\pi\overline{r}}{\lambda}\,\mathrm{Im}\left(\frac{\hat{m}^2 - 1}{\hat{m}^2 + 2}\right) \tag{69}$$

Here, $\hat{m}$ is the complex refractive index of the condensed material. Thus, it can be assumed that absorption by the water aerosol in CO_2-GDL makes the main contribution to the laser radiation attenuation coefficient

$$\beta = S_L(Q' + Q'') \tag{70}$$

An estimate of the value of Q'' for a value of $\hat{m} = 1.22 - 0.061i$ [85] corresponding to the 10.6 μm wavelength shows that even for a high average mass density of aerosol ($\rho_L = 10^{-6}$ g/cm^3), which according to the data cited in Fig. 11, *b* is possible in CO_2-GDL with a large initial water vapor content ($p_s > 1$ atm), the absorption coefficient

$$\beta \simeq 0{,}98\, S_L\, \overline{r}/\lambda$$

does not exceed 10^{-3} cm^{-1} and, consequently, is significantly less than the gain coefficient $\alpha \simeq 5 \cdot 10^{-3}$ cm^{-1} at the 00^01-10^00 transition that is typical for a homogeneous medium. Nevertheless, the effect of the optical channel of heterogeneous losses on the operation of CO_2-GDL can be quite significant. Actually, from calculating the efficiency of converting energy into laser radiation [15] performed assuming optimum mirror transmission and the absence of other loss sources in the resonator (β_0 = 0), it follows that when β = 0.1α, for example, the fraction of laser radiation absorbed by water condensate particles,

$$\delta = 2\sqrt{\beta/\alpha} - \beta/\alpha. \tag{71}$$

is only 50%.

Another potentially more effective relaxation channel is that of heterogeneous losses in CO_2-GDL. In the previous section, the basic physical mechanisms of the catalytic action of aerosol particles on the vibrational relaxation processes of gas molecules were examined. We shall focus in more detail on the features of the relaxation activity of water condensate particles in CO_2-GDL. Recall that the most probable upper CO_2 laser level (00^01) relaxation channel is the process of intermolecular nonresonant quantum exchange between the antisymmetric ν_3 and symmetric ν_1, ν_2 CO_2 vibrational modes. When CO_2 molecules collide with gas-phase water molecules, the exchange probability is 10^{-3}. If this process occurs on the surface of water condensate particles, then, as calculations show for the collisional deactivation mechanism done using Eq. (28) with the corresponding parameters of the interaction potential, its probability P^s turns out to be on the order of 1. With such a large probability as calculated for a single collision with water condensate particles, surface relaxation of the symmetric vibrations of CO_2 molecules is also realized, the level of excitation of which determines the population of the lower laser levels 10^00 and 02^00. As for surface deactivation of nitrogen molecules, which is caused by nonresonant quantum exchange with rapidly relaxing bending vibrational mode of H_2O, it occurs much less efficiently and is characterized by a probability of $P^s \sim 10^{-3}$.

This probability difference of surface vibrational deactivation of CO_2 and N_2 molecules suggests that relaxation of nitrogen molecule's vibrational energy, as in heterogeneous $CO_2(1)$-$N_2(2)$ mixtures containing water aerosols (H_2O), is mostly a gradual process due to resonant transfer of vibrational excitation by CO_2 molecules

$$N_2^* + CO_2 \xrightarrow{k_{21}} N_2 + CO_2^*(\nu_3)$$

and subsequent deactivation of antisymmetric vibrational modes during collisions of CO_2 molecules with condensate $(H_2O)'$ particles and water vapor molecules $(H_2O)''$:

$$CO_2^*(\nu_3) + (H_2O) \xrightarrow{k_1} CO_2 + (H_2O).$$

From the kinetics equations describing this scheme of the processes, it follows that in heterogeneous mixtures with a high relative concentration of nitrogen ($\kappa_2 >> \kappa_1$), for which a quasi-steady distribution of quanta is possible between the antisymmetric CO_2 mode and an N_2 vibration, the relaxation of the vibrational energy of the N_2 molecule

$$de_2/dt = -w(e_2 - \bar{e}_2) \tag{72}$$

happens at the rate

$$w = \frac{\kappa_1}{\kappa_2} \frac{\Gamma_1\Gamma_{12}}{\Gamma_1 + \Gamma_{12}}, \tag{73}$$

where Γ_1 and Γ_{12} are the rates of VT-deactivation of CO_2 molecules and resonant vibrational exchange with N_2, respectively.

Thus, condensate particles, by increasing the rate Γ_1 of direct VT-relaxation of CO_2 molecules, thereby accelerate vibrational relaxation of nitrogen molecules as well. As is evident from Eq. (73) and the expression for the surface component of Γ_1:

$$\Gamma_1^{(s)} = \kappa n q_L k_1^{(s)} = P_1^s \bar{v} S_L \left[1 + d/\bar{r} + (d/2\bar{r})^2\right], \tag{74}$$

the rate of relaxation of N_2 molecules turns out to be higher the greater the share of carbon dioxide gas in the mixture and the higher the dispersion of the water condensate are. A calculation of Γ_1 for a heterogeneous mixture of CO_2:N_2:$(H_2O)''$:$(H_2O)'$ = 1:8:0.5:0.5 with aerosol parameters of $S_L = 10\ cm^{-1}$, $\bar{r} = 10^{-7}$ cm, typical for a CO_2-GDL with a high initial water vapor content (see Fig. 11, *b*) shows that, when the heterogeneous channel of CO_2 molecular relaxation appears, the typical decay length of molecular nitrogen vibrations in a flow u/w for $P_1^s = 1$ is shortened by a factor of 5 and is just 3 cm. It is obvious that, with such a drop in the capacity of nitrogen molecules to preserve their vibrational excitation in a heterogeneous mixture, it is difficult to avoid significant energy losses in the gain medium when supplying this mixture to the CO_2-GDL resonator.

In addition to the increase in the relaxation loss power, surface deactivation of CO_2 molecules also causes additional optical losses. They are caused by the decrease in the gain coefficient of the active medium arising due to accelerating vibrational relaxation of CO_2 and N_2 molecules by water condensate particles. For a high relative concentration of nitrogen in the mixture, the amount of drop in α at small distances

$$u/w \gg x \geqslant u/(\Gamma_1 + \Gamma_{12})$$

from the condensation front is primarily determined by that change in the energy of the antisymmetric CO_2 mode, which is related to the transition from a quasi-steady vibrational distribution between CO_2 and N_2 molecules

$$e_1 \simeq \tilde{e}_1 = (\Gamma_1 \bar{e}_1 + \Gamma_{12} e_2)/(\Gamma_1 + \Gamma_{12}),$$

of the corresponding homogeneous ($q_L = 0$) relaxation regime, to the heterogeneous ($q_L \neq 0$) regime. It is not hard to see that, in this case, the decrease in the gain coefficient in comparison with homogeneous conditions will be characterized by the quantity

$$[\alpha(0) - \alpha(q_L)]/\alpha(0) = [\Gamma_1(q_L) - \Gamma_1(0)]/[\Gamma_1(0) + \Gamma_{12}] \; . \tag{75}$$

A calculation done using Eq. (75) shows that for the example examined above $\Delta\alpha/\alpha(0) \simeq 0.5$, and the additional optical losses, caused by this decrease in α in the biphase mixture flow region, for $\beta/\alpha(0) = 0.1$ reach 20%. In all, the optical losses along the heterogeneous channel are not 50%, as would be the case in the absence of relaxation activity of the aerosol particles, but 70%.

Figure 12 shows the axial gain coefficient α and the vibrational $CO_2(\nu_3)$ and N_2 temperature distributions, as calculated allowing for heterogeneous molecular relaxation in a jet of CO_2-N_2-H_2O as applied to the experimental conditions of Ref. [81]. The dot-dashed line shows the path of these curves that were calculated in the absence of water vapor condensation, and the points show the experimental values of α. A calculation of the population inversion at the CO_2 transition 00^01-10^00 was done using the method described in Section 1. The calculated data on the condensate, formed under conditions corresponding to these experiments (see Fig. 12, curve *2*), were used as the initial data for this population inversion calculation. The surface deactivation probabilities of CO_2 and N_2 molecules when colliding with water aerosol particles were taken to be 1 and 10^{-3}, respectively. In the precondensation region of the supersonic flowstream, where only water dimers are present in the greatest concentration, the effect of the aerosol on the kinetics of relaxation processes was neglected.

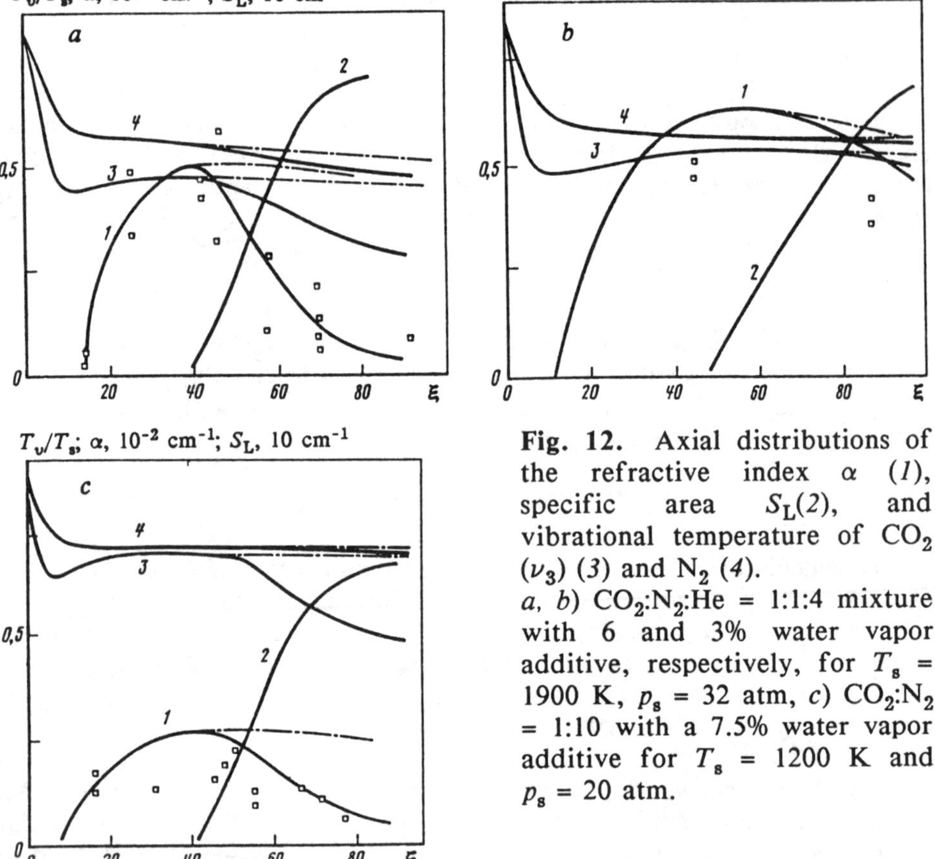

Fig. 12. Axial distributions of the refractive index α (*1*), specific area S_L(*2*), and vibrational temperature of CO_2 (ν_3) (*3*) and N_2 (*4*).
a, b) CO_2:N_2:He = 1:1:4 mixture with 6 and 3% water vapor additive, respectively, for T_s = 1900 K, p_s = 32 atm, *c*) CO_2:N_2 = 1:10 with a 7.5% water vapor additive for T_s = 1200 K and p_s = 20 atm.

As is evident from Fig. 12, the calculation is in satisfactory agreement with experiment not only in terms of the behavior of the function, but also in terms of the absolute value of α in the water vapor condensate region. The most distinct relaxation activity of aerosol particles appears in experiments with a 6% water vapor additive, where a sharp drop in α is caused by the formation of a large number of highly disperse water condensate (see Fig. 12, *a*). In this sense, indicative are experiments with a mixture in which 7.5% water vapor was added and which contains a large amount of nitrogen (90%), capable of restraining the heterogeneous relaxation of CO_2 molecules to a significant degree by *VV'*-exchange (see Fig. 12, *c*). According to the calculations, the aerosol parameters for the mixture should be the same as for a mixture with a 6% H_2O additive. However, the drop in α in the condensate region, as would be expected on the basis of Eq. (75), turns out to be less. This once again attests to the dominant role of the relaxation activity of aerosol particles, as opposed to their optical activity, since in the latter case the variation of the gain coefficient in both series of experiments would be identical and extremely small in value $\Delta\alpha < 10^{-3}$ cm^{-3}. As for the experiments with a 3% water vapor additive (see Fig. 12, *b*), as a consequence of the significant stretching of the condensation front, the specific area and mass fraction of the water aerosol particles in the observation region turn out to be almost an order of magnitude less than under the conditions examined above. Therefore, the relaxation effect of the water vapor condensate is slight here and can be insignificant in experiments to measure α.

3.3. Condensation of Carbon Dioxide Gas and Heterogeneous Losses in a CO_2 Coupled-Mode Gas-Dynamic Laser

When discussing the possibility of producing GDL based on coupled modes, the need to allow for heterogeneous losses during the condensation of carbon dioxide gas was mentioned. The specifics of this problem consist in the fact that the component condensing in this case, playing not an assisting role, as for example water vapor in the usual type of CO_2-GDL, but the primary role, is the radiant energy carrier. This fact complicates making a detailed analysis of the heterogeneous losses in coupled mode CO_2-GDL, since in the presence of vibrational inhomogeneity intrinsic to the energy carrier, heterogeneous relaxation of CO_2 molecules can also exert a significant influence on the kinetics of carbon dioxide gas condensation.

Without stopping to consider in detail those effects caused by the vibrational nonequilibrium of a supersaturated gas, we note that energy exchange between molecules' vibrational and translational degrees of freedom should inhibit the condensation process [95]. A direct cause of this can also be the decrease in the sticking probability of carbon dioxide molecules that is related to surface deactivation, as well as the nonisothermal behavior in the gas-condensate system that is caused by the vibrational relaxation of molecules in drops. It does not seem possible to correctly account for these effects in the framework of classical condensation theory. Therefore, we shall confine ourselves to exposing the qualitative features of the effect of carbon dioxide gas condensation on CO_2-GDL operation.

Figures 13 and 14 show a few results of modeling of carbon dioxide gas condensation as a CO_2-Ar mixture flows out of a plane slit into a vacuum. The vibrational state of the CO_2 molecules while condensing in an adiabatically expanding gas was assumed to be "quenched". In conjunction with this, the calculation scheme for the formation and growth kinetics of the condensed phase was taken to be the same as for the water vapor case. When choosing the thermodynamic quantities, we assumed that the CO_2 condensate was in a supercooled liquid state. Therefore, the saturation line at temperatures below the triple point (T_{tr} = 217 K) was modified and was given by the function [96, 97]

$$P_{\infty}(\text{atm}) = \exp(11.45 - 2127/T)$$

The value of σ_{∞} was found by extrapolating the function [97, 98]

$$\sigma_{\infty}(\text{J/cm}^2) = (55 - 0.184T)\cdot 10^{-7}$$

to the low temperature region. The value α = 2.5 Å was used as the correction to the nucleation center size. The drop's growth rate was calculated in the thermal equilibrium approximation in the gas-condensate system, and the molecular sticking probability was assumed equal to unity.

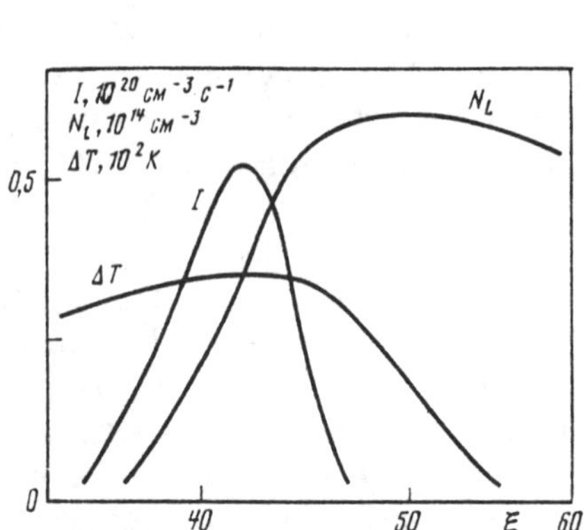

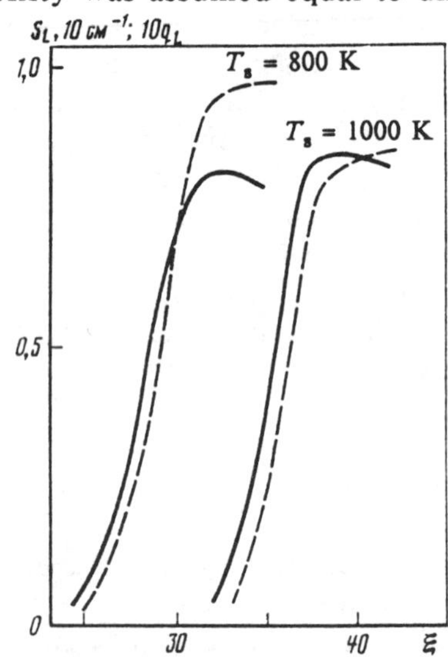

Fig. 13. Axial distributions of the rate of nucleation center formation I, condensate particle concentration N_L, and supercooling depth ΔT of carbon dioxide gas in a CO_2:Ar = 1:1 jet for T_s = 1000 K, p_s = 10 atm.

Fig. 14. Axial distributions of the specific area S_L (solid curves) and mass fraction q_L (dashed curves) of CO_2 condensate in a CO_2:Ar = 1:1 jet at different stagnation temperatures and $n_s = 1.45\cdot 10^{20}$ cm^{-3}.

As is evident from Fig. 13, carbon dioxide gas in this case can condense intensely only for sufficiently deep supercooling, when it reaches ΔT = 30-35 K, and the temperature of the mixture turns out to be below 100 K. The condensate so formed is characterized by the mass fraction q_L, which does not exceed the value

$$q_L = (\Delta T)_{max}/\kappa L(\gamma - 1),$$

necessary to completely remove the supersaturation attained in the flowstream. It is important that, by virtue of the relatively weak dependence of $(\Delta T)_{max}$ on the initial conditions, the mass fraction and specific area of the aerosol particles behind the condensation front are almost constant over the examined range of stagnation temperatures and pressures (see Fig. 14). Only the position and curvature of the condensate front vary: with an increase in p_s or decrease in T_s, it is moved toward the slit and is narrowed.

The condensation front being close to the slit limits the capability of using low-temperature thermal pumping in coupled mode CO_2-GDL. This is mostly caused by the difficulty of providing the vibrational nonequilibrium of CO_2 molecules required for lasing at the 03^10-10^00 transition in the biphase mixture flow region. The catalytic effect of condensation, leading to a drop in the symmetric mode temperature T_{12}, is not only important here, but also the thermal effect, which causes an increase in the gas temperature.

Figure 15 shows the axial distributions of the aforementioned temperatures and relative population inversion $\Delta N_v/N'_v$, calculated for the 03^10-10^00 transition allowing for gas-dynamic (thermal) and relaxation activity of carbon dioxide gas condensation in the flowstream. The heterogeneous relaxation of the CO_2 symmetric mode energy was calculated in the approximation of a quasi-steady distribution of quanta between CO_2 molecules in the gaseous and condensed phases using Eq. (54). To estimate the upper boundary of the heterogeneous losses, we assumed that the decay of the CO_2 symmetric vibrations in the aerosol particles takes place just as fast as in a liquid, and is characterized by a rate of 10^6 s^{-1}, which corresponds to the vibrational-translational relaxation of CO_2 (ν_3) for binary collisions of molecules in a gaseous medium with the density of the condensed phase. From Fig. 15 it is clear that, in this case, carbon dioxide gas condensation can cause a rather sharp drop in the symmetric vibration temperature. This, and the stabilization of the gas temperature at the level $T \simeq T_\infty$, leads to the population inversion disappearing at the laser transition behind the condensation front in the flowstream. In this respect, the stagnation parameters are critical, for which carbon dioxide gas begins to condense intensely in the region of the jet, where the temperature $T = T_{\infty} - (\Delta T)_{max}$ turns out to be the maximum attainable ($T = T_{12}/5.4$ (see condition (23)). The corresponding maximum values of the initial pressures p_s^{max}, calculated for a CO_2:Ar = 1:1 mixture in the assumption of instantaneous quenching of the symmetric vibrations, are shown in Fig. 16. For $p_s \geq p_s^{max}$, the condensation of carbon dioxide makes it impossible for a population inversion to form at the 03^10-10^00 transition in the flowstream.

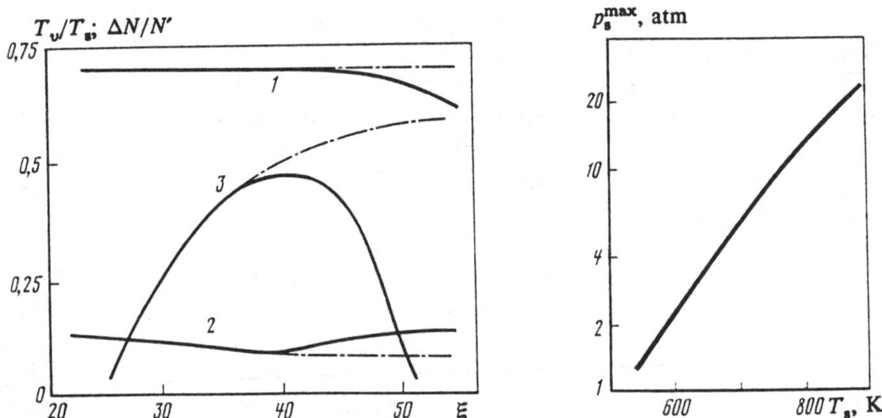

Fig. 15. Axial distributions of the vibrational (*1*), gas (*2*) temperatures and the relative vibrational population inversion (*3*) at the 03^10-10^00 transition in a jet of condensing carbon dioxide gas (solid curves) and in the absence of condensate (dot-dashed line).
CO_2:Ar = 1:1 mixture at T_s = 1000 K, p_s = 10 atm.

Fig. 16. Maximum achievable stagnation pressure for a CO_2-GDL based on a mixture of CO_2:Ar = 1:1.

In conclusion, we shall estimate the effect of the optical heterogeneous loss channel. For simplicity we shall assume that the interaction of laser radiation with condensate molecules is resonant, i.e., there is no frequency shift of the symmetric vibrations when the molecules make the transition from the gaseous to the condensed phase. Then, as a function of the degree of vibrational nonequilibrium of the CO_2 condensate molecules, the latter can both attenuate and amplify the radiation generated by the carbon dioxide gas molecules. If one is interested in absorption, which is possible due to the strong attenuation of the intrinsic molecular vibrations in the liquid phase and the nonisothermal behavior in the gas-condensate system, it should be insignificant for two reasons. These are, firstly, the assumptions made above regarding the vibrational and thermal states of the condensate molecules and, secondly, the radiation line-width $\Delta\overline{\nu}$ of the condensate molecules being significantly greater than in the condensed phase. The latter is the dominant factor and determines the scale of the difference between the absorbing and amplifying efficiencies of the condensed and gas phase molecules, respectively. Assuming for CO_2 condensate molecules, T_{12} = 500 K, T = 120 K, A_{12} = 0.1 s^{-1}, $\Delta\overline{\nu}$ = 10 cm^{-1}, we obtain that, under the conditions of Fig. 15, the ratio of β/α in the biphase flow region of the jet is on the order of 10^{-4}, and the optical losses are 2%. As for the amplification of radiation by condensate molecules, which is possible with weak attenuation of the symmetric vibrations in the solid phase, as a similar calculation shows, it is extremely small on the background of amplification in the gas.

REFERENCES

1. N. G. Basov and A. N. Oraevskii, "Producing negative temperatures by the method of heating and cooling the system", ZH. EKSP. TEOR. FIZ., Vol 44 No 5 pp 1742-1745.

2. Pat. 3015072 (US), Thermal Maser, H. E. D. Scovil and E. O. Shulz-Dubois, 1961.

3. I. R. Hurle and A. Hertzberg, "Electronic population inversions by fluid-mechanical techniques", PHYS. FLUIDS, Vol 8 No 9 pp 1601-1607, 1965.

4. N. G. Basov, A. N. Oraevskii, and V. A. Sheglov, "Thermal methods for exciting lasers", ZH. TEKH. FIZ., Vol 37 No 2 pp 339-348, 1967.

5. V. K. Konyukhov and A. M. Prokhorov, "Population inversion after adiabatic expansion of a gaseous mixture", PIS'MA ZH. EKSP. TEOR. FIZ., Vol 3 No 11 pp 436-439, 1966.

6. D. M. Kuehn and D. J. Monson, "Experiments with a CO_2 gas-dynamic laser", APPL. PHYS. LETT., Vol 16 No 1 pp 48-50, 1970.

7. A. P. Dronov, A. S. D'yakov, E. M. Kudryavstev, and N. N. Sobolev, "Gas-dynamic CO_2-laser with a shock-tube-heated gain medium flowing through a nozzle", PIS'MA ZH. EKSP. TEOR. FIZ., Vol 11 No 11 pp 516-519, 1970.

8. V. K. Konyukhov, I. V. Matrosov, A. M. Prokhorov, et al., "Gas-dynamic cw laser based on a mixture of carbon dioxide gas, nitrogen, and water", PIS'MA ZH. EKSP. TEOR. FIZ., Vol 12 No 10 pp 461-466, 1970.

9. E. T. Gerry, "Gas-dynamic laser", LASER FOCUS, Vol 6 No 12 pp 27-32, 1970.

10. A. Kantrowitz, "Heat-capacity lag in gas-dynamics", J. CHEM. PHYS., Vol 14 No 12 pp 150-164, 1946.

11. I. R. Hurle, A. L. Russo, and J. G. Hall, "Spectroscopic studies of vibrational nonequilibrium in supersonic nozzle flow", J. CHEM. PHYS., Vol 40 No 8 pp 2076-2089, 1964.

12. B. K. Bronfin, L. R. Boedeker, and J. P. Cheyer, "Thermal laser excitation by mixing in a highly convective flow", APPL. PHYS. LETT., Vol 16 No 5 pp 214-216, 1970.

13. R. Borghi, A. P. Carrega, M. Charpenel, and J. P. E. Taran, "Supersonic mixing nozzle for gas-dynamic lasers", APPL. PHYS. LETT., Vol 22 No 12 pp 661-663, 1973.

14. V. H. Krosko, P. I. Soloukhin, and N. A. Formin, "Effect of the composition and temperature of a medium on the efficiency of thermal excitation of the population inversion by mixing in a supersonic flow", FIZ. GORENIYA VZRYVA, Vol 10 No 4 pp 473-485, 1974.

15. S. A. Losev, "Gas-Dynamic Lasers", New York, Springer-Verlag, 1981.

16. T. A. Cool, "MDI - the transfer chemical laser: a review of recent research", IEEE J. QUANT. ELECTR., Vol QE-9 No 1 pp 72-83, 1973.

17. D. J. Spencer, H. Mirels, and D. A. Duran, "Performance of a cw chemical laser with N_2 or He diluent", J. APPL. PHYS., Vol 43 No 3 pp 1151-1157, 1972.

18. A. E. Hill, Uniform electrical excitation of a large volume high-pressure near-sonic CO_2-N_2-He flowstream", APPL. PHYS. LETT., Vol 18 No 5 pp 194-196, 1971.

19. A. S. Biryukov, "Kinetics of physical processes in gas-dynamic lasers", TRUDY FIAN, Vol 83 pp 13-83, 1975.

20. Yu. I. Grin', V. M. Polyakov, and V. G. Testov, "Experimental investigation of a N_2O-N_2-He mixture gas-dynamic laser amplifier", PIS'MA ZH. EKSP. TEOR. FIZ., Vol 18 No 4 pp 260-262, 1973.

21. V. F. Gavrikov, A. P. Dronov, V. K. Orlov, and A. K. Piskunov, "Nitrous oxide gas-dynamic laser", KVANT. ELEKTRON., No 5 pp 119-120, 1973.

22. A. S. Biryukov, A. Yu. Volkov, A. I. Demin, et al., "Investigation of a gas-dynamic N_2O-laser", ZH. EKSP. TEOR. FIZ., Vol 68 No 5 pp 1664-1677, 1975.

23. V. F. Gavrikov, A. P. Dronov, V. K. Orlov, and A. K. Piskunov, "Carbon bisulfide gas-dynamic laser", PIS'MA ZH. EKSP. TEOR. FIZ., Vol 23 No 11 pp 649-650, 1976.

24. A. Yu. Volkov, A. I. Demin, V. N. Epikhin, and E. M. Kudryavtsev, "Potential to increase the efficiency of gas-dynamic lasers: a carbon bisulfide gas-dynamic laser", KVANT. ELEKTRON., Vol 3 No 8 pp 1833-1836, 1976.

25. N. G. Basov, V. K. Mikhailov, A. N. Oraevskii, and V. A. Sheglov, "Producing a population inversion of molecules in a supersonic flow of a binary gas in a Laval nozzle", ZH. TEKH. FIZ., Vol 38 No 12 p 2031, 1968.

26. N. I. Yushchenkova and Yu. A. Kalenov, "Chemical and vibrational relaxation in supersonic carbon dioxide flows", ZH. PRIKL. SPEKTROSK., Vol 11 No 3 pp 417-424, 1969.

27. A. S. Biryukov, B. F. Gordiets, and L. A. Shelepin, "Vibrational relaxation and inverted population of molecular levels under nonsteady conditions", ZH. EKSP. TEOR. FIZ., Vol 57 No 2 pp 585-599, 1969.

28. B. F. Gordiets, N. N. Sobolev, and L. A. Shelepin, "Kinetics of physical processes in CO_2-lasers", ZH. EKSP. TEOR. FIZ., Vol 53 No 11 pp 1822-1833, 1967.

29. V. K. Konyukhov, "Gas-dynamic cw CO_2-laser", TRUDY FIAN, Vol 113, pp 50-113, 1979.

30. J. D. Anderson, "Time-dependent analysis of population inversions in an expanding gas", PHYS. FLUIDS, Vol 13 No 8 pp 1983-1989, 1970.

31. J. Tulip and H. Seguin, "Gas-dynamic CO_2-laser pumped by combustion of hydrocarbons", J. APPL. PHYS., Vol 42 No 9 pp 3393-3398, 1971.

32. N. A. Generalov, G. I. Kozlov, and I. K. Selezneva, "Population inversion of CO_2 molecules in an expanding gas flow", ZH. PRIKL. MEKH. TEKH. FIZ., No 5 pp 24-34, 1971.

33. S. A. Losev, V. N. Makarov, V. A. Pavlov, and O. P. Shatalov, "Investigation of the processes in a gas-dynamic laser in a large diameter shock tube", FIZ. GORENIYA VZRYVA, Vol 9 No 4 pp 463-473, 1973.

34. E. M. Kudryavtsev and V. N. Faizulaev, "Formation of a population inversion a jet of gaseous mixture CO_2-N_2-H_2O expanding through a slit", ZH. PRIKL. MEKH. TEKH. FIZ., No 6 pp 25-31, 1973.

35. V. M. Shmelev, N. Ya. Vasilik, and A. D. Margolin, "Gas-dynamic laser based on a CO_2-N_2-CO-H_2O mixture", KVANT. ELEKTRON., Vol 1 No 8 pp 1711-1715, 1974.

36. A. I. Demin, E. M. Kudryavtsev, N. N. Sobolev, and V. N. Faizulaev, "Experimental investigation of the limiting water vapor content in a gas-dynamic laser based on CO_2-H_2O-N_2", KVANT. ELEKTRON., Vol 1 No 3 pp 528-533, 1974.

37. A. A. Likal'ter, "Relaxation of symmetric vibrational modes of CO_2 molecules", ZH. PRIKL. MEKH. TEKH. FIZ., No 3 pp 8-17, 1975.

38. C. E. Treanor, J. W. Rich, and R. G. Rehm, "Vibrational relaxation of anharmonic oscillators with exchange-dominated collisions", J. CHEM. PHYS., Vol 48 No 4 pp 1798-1807, 1968.

39. E. M. Kudryavtsev and V. N. Faizulaev, "Population inversion in a gas mixture jet containing carbon dioxide and expanding through a slit", KVANT. ELEKTRON., Vol 1 No 10 pp 2230-2238, 1974.

40. V. K. Konyukhov and V. N. Faizulaev, "Potential for making a gas-dynamic laser based on transitions between levels of coupled modes of CO_2", KVANT. ELEKTRON., Vol 5 No 12 pp 2620-2621, 1978.

41. V. K. Konyukhov and V. N. Faizulaev, "Potential for making a gas dynamic laser based on transitions between levels of coupled modes of CO_2", Preprint FIAN No 204, Moscow, 1978, 11 pp.

42. A. A. Vedeneev, A. Yu. Volkov, A. I. Demin, et al., "Gas-dynamic

laser with thermal pumping at transitions between the bending and symmetric modes of CO_2", PIS'MA ZH. TEKH. FIZ., Vol 4 No 11 pp 681-684, 1978.

43. N. M. Kuznetsov, "Vibrational quasi-equilibrium with rapid quantum exchange", ZH. EKSP. TEOR. FIZ., Vol 61 No 9 pp 950-955, 1971.

44. A. S. Biryukov and B. F. Gordiets, "Kinetic equations of the relaxation of vibrational energy in a mixture of monatomic gases", ZH. PRIKL. MEKH. TEKH. FIZ., No 6 pp 29-37, 1972.

45. B. F. Gordiets, A. I. Osipov, E. V. Stupochenko, and L. A. Shelepin, "Vibrational relaxation in gas and molecular lasers", USP. FIZ. NAUK, Vol 108 No 4 pp 655-699, 1972.

46. R. L. Taylor and S. Bitterman, "Survey of vibrational relaxation data for processes important to the CO_2-N_2 laser system", REV. MOD. PHYS., Vol 41 No 1 pp 26-47, 1969.

47. O. N. Katskova, I. N. Naumov, Yu. D. Shmyglevskii, and N. P. Shumilin, "Opit rascheta ploskikh i osesimmetrichnykh sverkhzvukovykh techenii metodom kharakteristick [Attempt to model planar and axially-symmetric supersonic flows by the method of characteristics]", VTs AN SSSR, Moscow, 1961, 72 pp.

48. A. C. G. Mitchell and M. W. Zemansky, "Resonance Radiation and Excited Atoms", New York, Cambridge, 1961.

49. V. V. Danilov, E'. P. Kruglyakov, and E. V. Shun'ko, "Measurement of the probability of the transition P(20) 001-100 of CO_2 and shock broadening in collisions of CO_2 with N_2 and He", ZH. PRIKL. MEKH. TEKH. FIZ., No 6 pp 24-28, 1972.

50. A. A. Likal'ter, "Laser based on the transitions between levels of coupled modes of CO_2", KVANT. ELEKTRON., Vol 2 No 11 pp 2399-2402, 1975.

51. C. Simpson, T. Chandler, and A. S. Strawson, "Vibrational relaxation in CO_2 and CO_2-Ar mixtures studied using a shock tube and laser-schlieren technique", J. CHEM. PHYS., Vol 51 No 5 pp 2214-2220, 1969.

52. H. E. Bass, "Vibrational relaxation in CO_2/O_2 mixtures", J. CHEM. PHYS., Vol 58 No 11 pp 4783-4786, 1973.

53. K. Schafer and G. G. Grau, "Thermiche Akkomodation und die Geschwindigkeit der Energieübertragung von CS_2 an Platin in Abhangigkeit von der Temperatur", ZTSCHR. ELECKTROCHEM., Vol 53 No 4 pp 203-210, 1952.

54. K. Schafer, "Energieübergungsmechanismus und Reactionsgeschwindigkeit an metallischen", ZTSCHR. ELECKTROCHEM., Vol 56 No 4 pp 398-405, 1955.

55. M. Kovacs, D. R. Rao, and A. Javan, "Study of diffusion and wall de-excitation probability of 001 state in CO_2", J. CHEM. PHYS., Vol 48 No 7 pp 3339-3340, 1968.

56. A. Black, H. Wise, S. Schechter, et al., "Measurements of vibrationally excited molecules by Raman scattering II: Surface deactivation of vibrationally excited N_2", J. CHEM. PHYS., Vol 60 No 9 pp 3526-3636, 1974.

57. L. Doyennett, M. Margottin-Maclou, H. Guerguen, et al., "Temperature dependence of the diffusion and accommodation coefficients in nitrous oxide and carbon dioxide into the vibrational level", J. CHEM. PHYS., Vol 60 No 2 pp 697-702, 1974.

58. Yu. M. Gershenzon, V. I. Egorov, and V. B. Rozenshtein, "Determining the accommodation coefficient of the vibrational energy of nitrogen molecules on the surface of molybdenum glass", KHIM. VYS. ENERG., Vol 7 No 6 pp 533-536, 1973.

59. S. A. Kovalevskii, "Methods for studying heterogeneous relaxation of vibrational energy of molecules", PROBLEMY KINETIKI I KATALIZA, Vol 17 pp 23-29, 1978.

60. Yu. M. Gershenzon, V. B. Rozenshtein, A. I. Spasskii, and A. M. Kogan, "Reactions of remnants of active particles in glass in flowstream conditions", DOKL. AKAD. NAUK SSSR, Vol 205 No 4 pp 871-873, 1972.

61. Yu. M. Gershenzon, V. B. Rozenshtein, and S. Ya. Umanskii, "Heterogeneous vibrational relaxation of N_2, CO_2, and N_2O molecules", PROBLEMY KINETIKI I KATALIZA, Vol 17 pp 36-59, 1978.

62. K. S. Gochelashvili, N. V. Karlov, A. I. Ovchenkov, et al., "Methods of selective heterogeneous separation of vibrationally excited molecules", ZH. EKSP. TEOR. FIZ., Vol 70 No 2 pp 531-542, 1976.

63. N. V. Karlov and K. V. Shaitan, "Separation of molecules with different isotopic composition during adsorption", ZH. EKSP. TEOR. FIZ., Vol 71 No 8 pp 464-471, 1976.

64. V. K. Konyukhov and A. M. Prokhorov, "Potential for making an adsorption-gas-dynamic laser", PIS'MA ZH. EKSP. TEOR. FIZ., Vol 13 No 8 pp 216-217, 1971.

65. A. S. Biryukov, V. M. Marchenko, and A. M. Prokhorov, "Population inversion of vibrational levels when mixing flows of nonequilibrium nitrogen and carbon dioxide aerosols", ZH. EKSP. TEOR. FIZ., Vol 71 No 11 pp 1726-1732, 1976.

66. V. M. Kuznetsov and M. M. Kuznetsov, "One gas-dynamic model of biphase flow with strong level nonequilibrium", TRUDY TsAGI, No 1932 pp 116-124, 1978.

67. V. K. Konyukhov and V. N. Faizulaev, "Kinetics of vibrational relaxation of molecules in a gas-aerosol system and biphase media lasers", KVANT. ELEKTRON., Vol 5 No 7 pp 1497-1498, 1978.

68. V. K. Konyukhov and V. N. Faizulaev, "Effect of condensation of the rates of relaxation processes in gas-dynamic lasers", KVANT. ELEKTRON., Vol 1 No 12 pp 2623-2625, 1974.

69. V. N. Faizulaev, "O relaksatsii kolebatel'noi energii molekul v geterogennykh smesyakh [Relaxation of vibrational molecular energy in heterogeneous media]", Preprint FIAN No 128, Moscow, 1979, 15 pp.

70. T. F. Hunter, "Vibrational energy transfer at gas-solid interfaces", J. CHEM. PHYS., Vol 51 No 6 pp 2641-2647, 1969.

71. E. E. Nikitin, "Teoriya atomno-moleculyarnykh protsessov v gazakh [Theory of Atomic and Molecular Processes in Gases]", Khimiya, Moscow, 1970, 456 pp.

72. R. N. Schwartz, Z. I. Slawsky, and K. F. Herzfeld, "Calculation of vibrational relaxation times", J. CHEM. PHYS., Vol 20 No 10 pp 1591-1598, 1952.

73. L. D. Landau and E. M. Lifshits, "Quantum Mechanics", New York, Pergamon, 1981.

74. J. Frenkel, "Kinetic Theory of Liquids", Oxford Univ. Press, London, 1946.

75. E. V. Stupochenko, S. A. Losev, and A. I. Osipov, "Relaksatsionnye protsessy v udarnykh volnakh [Relaxation Processes in Shock Waves]", Nauka, Moscow, 1965, 484 pp.

76. W. M. Madigosky and T. A. Litowitz, "Mean free path and ultrasonic vibrational relaxation in liquid and dense gases", J. CHEM. PHYS., Vol 34 No 2 pp 489-497, 1961.

77. B. F. Gordiets, Sh. S. Mamedov, and L. A. Shelepin, "Vibrational relaxation and lasers based on intramolecular transitions in liquid and molecular crystals", ZH. EKSP. TEOR. FIZ., Vol 69 No 8 pp 467-476, 1975.

78. W. F. Calaway and G. E. Ewing, "Vibrational relaxation of small molecules in the liquid phase: liquid nitrogen doped with O_2, CO, and CH_4", J. CHEM. PHYS., Vol 63 pp 2842-2844, 1975.

79. R. Merrey, "Vvedenie v yadernuyu tekhniku [Introduction to Nuclear Technology]", Moscow, Izd-vo inostr. lit., 1955.

80. V. I. Alferov, A. S. Biryukov, E. A. Bozhkova, et al., "Investigation of the interaction of hypersonic flows of nonequilibrium air with carbon dioxide aerosols", KVANT. ELEKTRON., Vol 6 No 7 pp 1746-1755, 1979.

81. A. I. Demin, E. M. Kudryavtsev, N. N. Sobolev, et al., "Effect of condensation of water vapor on the operation of a carbon dioxide gas-dynamic laser", KVANT. ELEKTRON., Vol 1 No 3 pp 706-708, 1974.

82. V. K. Konyukhov and V. N. Faizulaev, "Kondensatsiya parov vody i relaksatsionnye protsessy v gazodinamicheskogo CO_2 lazera [Condensation of water vapor and relaxation processes in a gas-dynamic CO_2 laser]", in: "Tez. dokl. II Vsesoyuz. simpoz. po fizike gazovykh lazerov [Proceedings of the II All-Union Symposium on the Physics of Gas Lasers] (16 July 1975, Novosibirsk), Moscow, 1975, p 131.

83. V. K. Konyukhov and V. N. Faizulaev, "Condensation of water vapor and relaxation processes in a gas-dynamic CO_2 laser", KVANT. ELEKTRON., Vol 5 No 3 pp 515-520, 1978.

84. A. B. Britan and A. M. Kortsenshtein, "Condensation of water vapor in a gas-dynamic CO_2 laser", KVANT. ELEKTRON., Vol 2 No 11 pp 2536-2537, 1975.

85. V. M. Shmelev, A. L. Margolin, "Heterogeneous processes in gas-

dynamic lasers", KVANT. ELEKTRON., Vol 2 No 11 pp 1729-1755, 1975.

86. H. G. Stever, "Phenomenon of condensation during high-speed flow", in: "Fundamentals of Gas Dynamics" (H. W. Emmons, editor.), Princeton, Princeton Univ. Press, 1958.

87. J. M. Calo and J. H. Brown, "The calculation of equilibrium mole fractions of polar-polar, nonpolar-polar, and ion dimers", J. CHEM. PHYS., Vol 61 No 10 pp 3931-3941, 1974.

88. A. A. Viktorova and S. A. Zhevakin, "Rotational spectrum of water vapor dimer", DOKL. AKAD. NAUK SSSR, Vol 194 No 2 pp 291-294, 1970.

89. A. Kantrowitz, "Nucleation in very rapid vapor expansions", J. CHEM. PHYS., Vol 19 No 9 pp 1097-1100, 1951.

90. W. G. Courtney, "Nonsteady state nucleation", J. CHEM. PHYS., Vol, 36 No 8 pp 2009-2017, 1962.

91. R. P. Andres and M. Boudart, "Time lag in multistate kinetics: nucleation", J. CHEM. PHYS. Vol 42 No 6 pp 2057-2064, 1965.

92. P. G. Hill, "Condensation of water vapor during supersonic expansion in nozzles", J. FLUID MECH., Vol 25 No 3 pp 593-620, 1966.

93. L. M. Davydov, "Investigation of nonequilibrium condensation in supersonic nozzles and jets", IZV. AKAD. NAUK SSSR MZhG, No 3, pp 66-73, 1971.

94. H. C. van de Hulst, "Light Scattering by Small Particles", New York, Wiley, 1957.

95. V. K. Konyukhov and V. N. Faizulaev, "Kinetic model of condensation of a supersaturated gas", ZH. PRIKL. MEKH. TEKH. FIZ., No 3 pp 41-47, 1977.

96. K. M. Duff and P. G. Hill, "Condensation of carbon dioxide in supersonic nozzles", in: "Proceedings of Heat Transfer and Fluid Mech. Inst.", Stanford Univ. Press, Stanford, 1966, pp 268-275.

97. P. A. Skovorodko, "Vliyanie gomogennoi kondensatsii v svobodnoi strue na intensivnost' molekulyarnogo puchka [Effect of homogeneous condensation in a free jet on the intensity of a molecular beam]", in: "Nekotorye zadachi gidrodinamiki i teploobmena [Several Problems in

Hydrodynamics and Heat Exchange]", Novosibirsk, 1976, pp 106-114.

98. M. P. Vukalovich and V. V. Altunin, "Teplofizicheskie svoistva dvuokisi ugleroda [Thermophysical Properties of Carbon Dioxide]", Atomizdat, Moscow, 1965, 455 pp.

CONTINUOUS GAS-DYNAMIC LASER BASED ON THE COUPLED MODES OF MOLECULAR CO_2

V. D. Logvinenko

0. Introduction

In 1978, the first reports [1, 2] appeared on the potential for a new type of lasing action in gas-dynamic lasers (GDL) based on molecular CO_2. Two interesting features of the new type of GDL attracted the attention of researchers: high specific energy reserve in comparison with traditional GDL based on a CO_2-N_2-H_2O mixture and the potential for operating at a significantly lower stagnation temperature than traditional GDL. Lasing is possible in the new type of GDL at several transitions of CO_2 molecules in the 16-22 μm range, which widens the selection of sources of intense coherent infrared radiation.

Likal'ter first indicated that it was possible to produce a population inversion between coupled mode (CM) levels (Fig. 1). He established [3] that there is a Treanor type quasi-steady coupled mode distribution:

$$n_{v_1 v_2^l 0} = n_{00^0 0} \exp[-v\hbar\omega_2/T_v - \Delta_{v_1 v_2^l 0}/T],$$

$$\Delta = \Delta_{v_1 v_2^l 0} = E_{v_1 v_2^l 0} - v\hbar\omega_2,$$

where $\hbar\omega_2$ is the vibrational quantum of the bending mode, T is the gas temperature, T_v is the vibrational temperature of the symmetric and bending vibrations with vibrational quantum numbers ν_1 and ν_2^l, respectively, $v = 2v_1 + v_2^l$, and Δ is the shift of the level relative to the corresponding level in the spectrum of the harmonic approximation. For a sufficiently high vibrational temperature T_v and low gas temperature T, the distribution has a notch in it (Fig. 2), which makes it possible to produce a population inversion over an entire set of transitions between coupled mode levels. The population inversion is formed when satisfying the condition

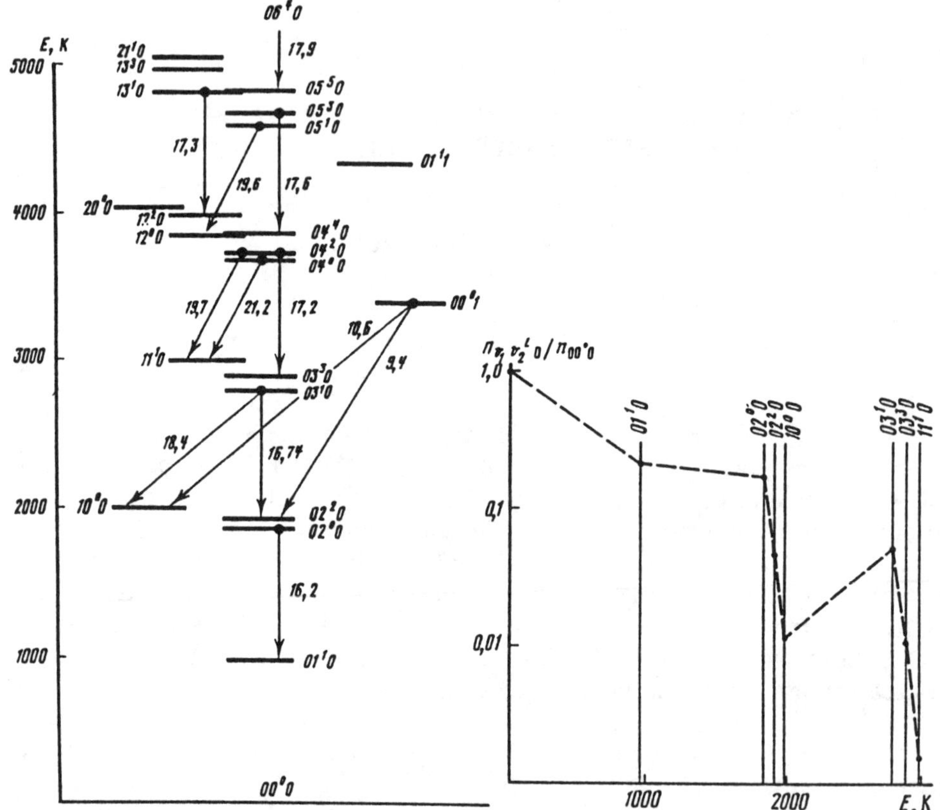

Fig. 1. Diagram of molecular CO_2 vibrational levels. The arrows denote transitions at which lasing action has been produced to date. The numbers by the arrows denote the wavelengths of the transitions in μm.

Fig. 2. Vibrational level population distribution at T_v = 560 K, T = 55 K.

$$\frac{T_v}{T} \geqslant \frac{1}{1-(E_{v+1,\beta}-E_{v,\beta})/\hbar\omega_2},$$

where $E_{v,\beta}$ is the energy of the βth component of the multiplet with vibrational quantum number v. The inversion condition is satisfied most simply for the 03^10-10^00 transition with an emission wavelength of λ = 18.4 μm:

$T_\upsilon/T \gtrsim 5{,}4$.

The mechanism for producing the population inversion between coupled mode levels is laid out in more detail in the first article of this volume. Likal'ter [4] examined a scheme for a flow-through gas-discharge laser. A shortcoming of the proposed scheme was the requirement for strongly diluting the CO_2 with argon (CO_2:Ar = 1:50) to provide a low gas temperature (T = 150 K).

It is much easier to provide the required population inversion when using gas-dynamic thermal pumping, which was independently suggested in Refs. [1, 2]. In a theoretical paper [1], it was reported that at the optimum gas mixture heating temperature and optimum dilution of CO_2:Ar = 1:1, the efficiency of converting thermal energy into radiation approaches 5-12%, and the gain coefficient can reach 2-3 m^{-1}. It was pointed out there that the depth of cooling the mixture can be easily varied over a wide range by changing the adiabatic index and by varying the dilution by the inert gas. The authors of Ref. [2] reported experimentally observing lasing action with a wavelength of λ = 18.4 μm in a GDL designed around a shock tube.

Later studies on GDL operating on coupled modes of CO_2 molecules (CM CO_2) were done in several directions. Laser action was produced on a whole series of vibrational-rotational transitions between coupled mode levels in the 16-22 μm wavelength region (see Fig. 1). The measurements were performed on an apparatus with an electric arc preheated mixture operating pulsed [5-10]. A nozzle system was used with a small critical cross-section height (h = 0.01 cm), which permitted us to attain deep cooling and high values of T_υ/T.

In a series of works [7-16], studies were conducted into the composition of the working mixture and the nozzle shape. As was indicated above, the ratio of T_υ/T - the degree of cooling - is a characteristic quantity that determines the operation of a CM CO_2-GDL. The cooling depth can be varied by choosing different degrees of nozzle opening or by varying the adiabatic index γ of the working mixture. An inert gas can be added to CO_2 to increase the value of γ. Lasing was produced for the mixtures CO_2-Ar(Ne, Xe, He), CO_2-N_2-Ar(Ne, He), as well as for pure CO_2 and the mixture CO_2-N_2. Simultaneous stimulated emission was produced at wavelengths of 10.6 μm and 18.4 μm in a CO_2-N_2-He mixture [16, 17].

An investigation into the effect of water, hydrogen, and helium impurities on the operating efficiency of a CM CO_2-GDL was conducted in Ref. [18]. Note that even 0.01% water has a significant effect on the population of vibrational levels ν_1 and ν_2. The effect of back pressure was studied In Ref. [19]. In a series of theoretical studies, the issues of the efficiency of operation of CM CO_2-GDL [11, 12, 20-23] and the causes that lower it were examined. Isolated among these were the nonideal nature of the gas-dynamic flow [19, 23], impurities [18], condensation [1], and the flux of vibrational energy through the upper levels [24, 25]. A brief survey of the research up to 1983 is given in Ref. [26].

To date, lasing has been produced in CM CO_2-GDL in several variants: in a shock-tube facility [2, 11, 14, 15] and on an electric arc preheated mixture apparatus operating pulsed [5-10, 16, 17] and cw [27]. In Ref. [27], experiments were conducted on estimating the gain coefficient k_y using the "calibrated losses" technique, which were set by varying the transmission of the resonator's output mirror. The value of k_y was determined by the point where lasing stopped. According to these estimates, the output energy density was roughly 20 J/g, the efficiency was $\approx$ 3%, and the gain coefficient was $k_y \approx 0.3$ m^{-1}. Pulsed lasing occurred in an electronic discharge gas-dynamic laser with a wavelength of 16.3-16.9 μm, when the pumping was done by a glow-discharge, and the required cooling of the mixture was done by expansion in a supersonic flow [28].

A characteristic feature of the aforementioned CM CO_2-GDL devices is their high stagnation temperature ($\geq$ 1000 K). However, theoretical calculations shows that a population inversion is also possible at low initial temperatures (500-700 K). To verify the possibility of getting lasing action at coupled modes at low stagnation temperatures, a continuous wave coupled-mode gas-dynamic laser CM-GL device was constructed with the mixture preheated by an ohmic heater.

1. Description of the CM-GL Gas-Dynamic Apparatus

The CM-GL apparatus permits one to operate cw and to use different gas mixtures with an initial pressure of $P_0 \lesssim 760$ Torr. The working mixture is preheated by an ohmic heater and after flowing through the nozzle is collected in a vacuum chamber. The overall scheme of the apparatus is shown in Fig. 3. Structurally, the CM-GL consists of the following basic subassemblies: 1) gas-dynamic portion, 2) heater, 3) gaseous mixture inlet system, 4) heater power supply.

The gas dynamic portion of the apparatus creates a continuous supersonic flow of the gas mixture through the resonator region. After leaving the heater *10*, the gas mixture flows through rotational elbow *9* and collector *8* and impinges on the entrance of a nozzle system *7* (see Fig. 3). The collector serves to uniformly distribute the working mixture over the entire length of the nozzle. The gas pressure in the collector is measured by a standard manometer, the gas temperature was measured by chromel-alumel thermocouple transducers *11*, *13*, *14* at the exit from the heater and in the collector. One of the thermocouples (*13*) sits directly in front of the nozzle in the collector.

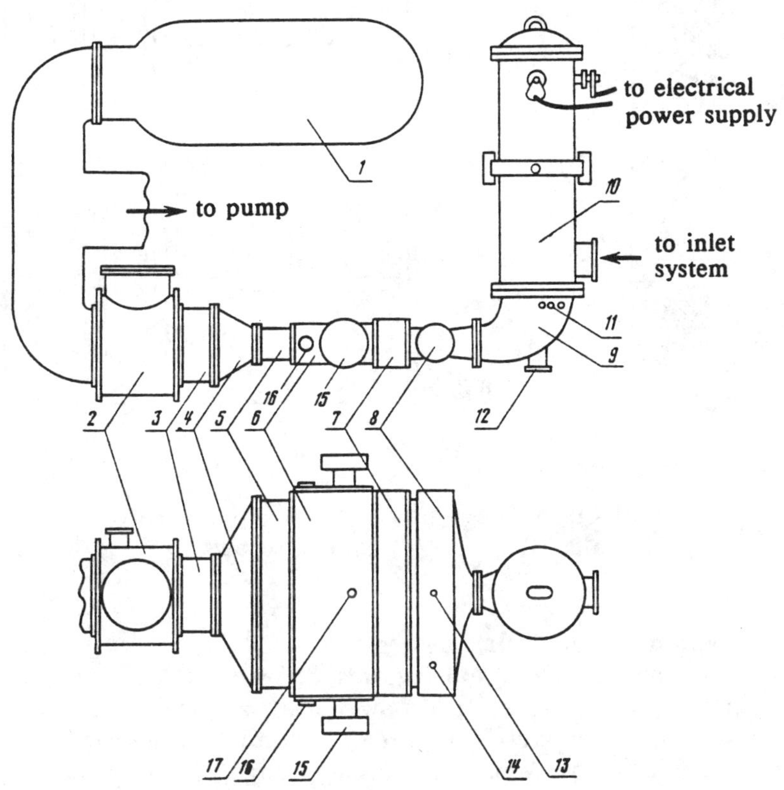

Fig. 3. Diagram of the CM-GL apparatus.
1) vacuum chamber, *2*) DU-500 vacuum valve, *3*) adaptor, *4*) expander, *5*) coupling, *6*) resonator system, *7*) nozzle system, *8*) collector, *9*) rotational elbow, *10*) heater, *11*, *13*, *14*) thermocouple transducer leads, *12*) inspection port for observing the Nichrome tube, *15*) mirrored tuning head, *16*) resonator inspection port, *17*) IKD6TDa-30 pressure transducer.

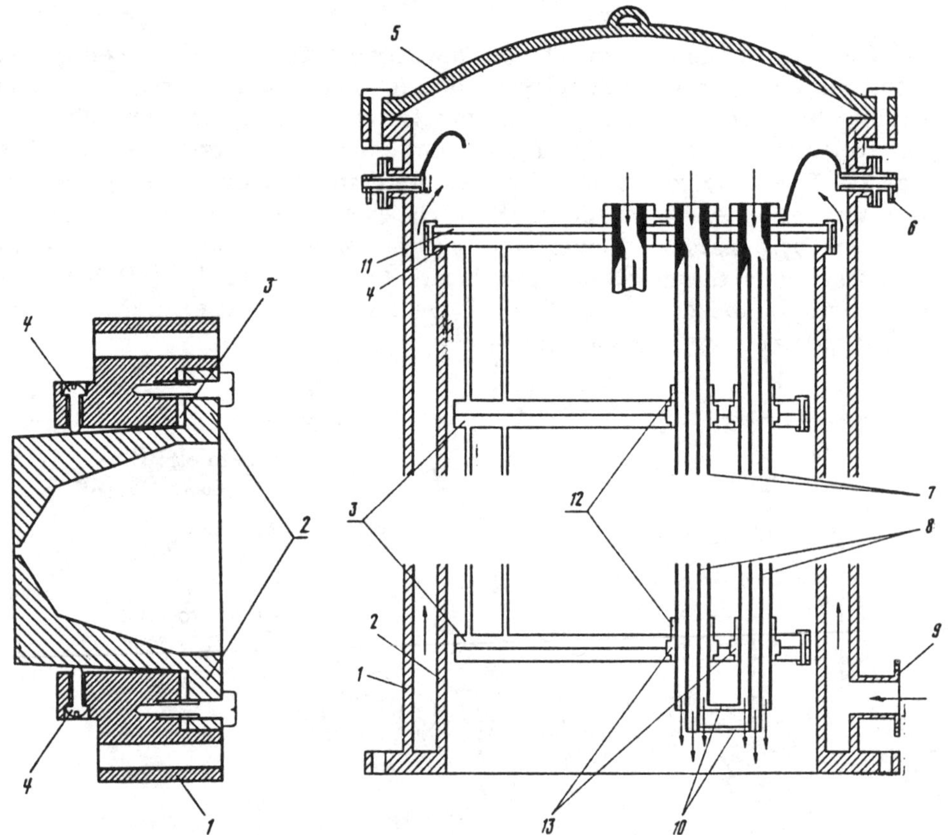

Fig. 4. Diagram of the nozzle system.
1) housing, *2*) slit walls, *3*) gasket, *4*) control screws.

Fig. 5. Diagram of the electric heater.
1, *2*) external and internal housing, respectively, *3*) support flanges, *4*) upper flange, *5*) cover, *6*) electric power lead, *7*, *8*) external and internal Nichrome tubes, respectively, *9*) flange for admitting the gas mixture, *10*) bulkheads, *11*) asbestos cement insulating flange, *12*) ceramic rings, *13*) ceramic collar; the arrows indicate the gas motion.

The nozzle apparatus consists of a horizontally oriented slit 70 cm long (Fig. 4). The height of the slit can be adjusted in the range from 0 to 1 mm. In the experiments that were conducted, the height of the slit was 0.4 mm (h_s = 0.2 mm). The cut of the slit was placed at the edge of the resonator block. The later is a rectangular chamber 40×90×100 mm in size. Freely expanding from the slit into the evacuated space of the resonator, the gas mixture is accelerated to supersonic speeds (Mach number $M \approx 9$), which ensures that a population inversion will be produced. There are flanges on the side walls of the resonator for joining the alignment nodes with the mirrors. In the experiments, polished copper mirrors and gold-coated glass mirrors 60 mm in diameter with a 2 mm diameter opening in the center were used to extract the radiation. The opening in the mirror was sealed with a thin Lavsan film in order to evacuate the resonator cavity. The length of the resonator was 1100 mm, the radius of curvature of the mirrors $R \approx 1200$ mm, i.e., the mirrors formed a nearly-confocal stable resonator. The distance from the slit of the nozzle to the optical axis of the resonator was roughly 20 mm. The gas pressure in the resonator was measured by an IKD6TDa-30 pressure transducer. After the resonator, the supersonic flow goes through the coupling *5* and impinges on the expander *4* and later goes through the adapter *3* and vacuum valve *2* into the vacuum chamber *1*, which is four cigar-shaped tanks of the same 720 m^3 capacity (see Fig. 3). This amount of vacuum capacity permits us to maintain stable gas-dynamic flow through the resonator for 1.5 min. The chamber was pumped down by two fore-vacuum pumps to an initial pressure on the order of 0.3 Torr.

1.2 Electric Heater

The heater was designed to produce the required temperature of the working mixture. The external form of the heater is shown in Fig. 5. The basis of the heater is an assembly of 16 double tubes (placed inside each other) made of grade KhN78T Nichrome, capable of achieving temperatures up to 1500 K. The external tube *7* has a diameter of 25 mm and a wall thickness of 1 mm, and the internal tube *8* has a diameter of 14 mm and a wall thickness of 1.5 mm. The wall thicknesses were chosen to make their linear ohmic impedances similar. The external and internal tubes were connected at the ferrule located above the heater. The ferrules were connected to the bulkheads, and every other pair of tubes below were also connected by bulkheads 10, such that all 16 pairs of tubes were connected in series with each other. The total electrical impedance of the heater in the cold state is 0.35 ohm. The tube assembly hangs vertically, supported on the upper flange *4*. To make the structure rigid, the tubes pass through five support flanges *3*. To electrically isolate the tubes from the body of the heater and the flanges *3*, *4*, a flange *11* made of asbestos cement and ceramic collars *13* were used. The collars are covered from above by ceramic rings *12* to eliminate the possibility of cinders and metallic dust incident from above blocking the area between the tubes and flanges *3*. The gas mixture enters the heater from the gas inlet system through flange *9*, supported from above between the external *1* and internal *2* bodies of the heater, and enters the cavity under the cover *5*. The gas is heated, passing downward between the internal external tubes and inside the internal tube, and exits through the lower flange of the heater into the gas-dynamic portion.

The two-tube construction - external and internal - was chosen to improve the thermal conductivity from the tube walls to the gas by decreasing the thickness of the heated gas layer. The total cross-sectional area for passing the gas between the tubes was 54.5 cm^2, which is roughly 20 times greater than the slit area of the nozzle system. When solving the thermal conductivity equation for a gas motion speed along the tubes of 15 m/s, it results that the gas acquires a temperature upon exiting from the heater that is ~80% of the tube temperature. The temperature of the external tubes is controlled by two chromel-alumel thermocouples placed in the upper and lower portions of the heater. The thermocouple readings were verified with pyrometer readings that measured the temperature of the internal tube. In the process of the tubes heating and cooling, the thermocouple readings lagged in comparison with the pyrometer readings by 2-3 s, i.e., during heating the thermocouples gave understated values and when cooling they gave overstated values. The difference in the thermocouple and pyrometer readings did not exceed 5% over the chosen operating regime. It is not possible to answer the question whether this reading difference is related to the inertia of the thermocouples or to a temperature difference between the internal and external tubes. In the process of heating, the temperature difference between the upper and lower portions of the external tube, judging from thermocouple data, can approach 100°.

During the experiments, the tubes warmed up to an operating temperature of 1000 K for roughly 20 min.

1.3 Gas-Mixture Inlet System

The inlet system serves to create a gaseous mixture of a given composition, store the mixture, and convey it to the heater inlet. A diagram of the inlet system is shown in Fig. 6. Gases from cylinders *1, 2* are fed in turn through the reducers *3* to the storage balloon *7*, which is usually made of a 5 m^3 capacity rubber meteorological weather balloon. A predefined amount of gas is controlled by the weight of the balloon.

The storage balloon is hung to the ceiling and rests on a column-tube *8*, which has numerous small openings with a total area on the order of about 100 cm^2. This construction provides, first of all, better gas mixing and, secondly, uniform collapse of the balloon's envelope to the tube during start-up of the device. The gas mixture in the balloon is always at atmospheric pressure. At start-up of the apparatus, the vacuum valve *5* is opened and atmospheric pressure compresses the rubber envelope of the balloon, feeding gas to the heater's input. A gauging collar *6* serves to control the gas flow rate and to vary the stagnation pressure in front of the slit. The instant start-up is completed, valve *5* is closed.

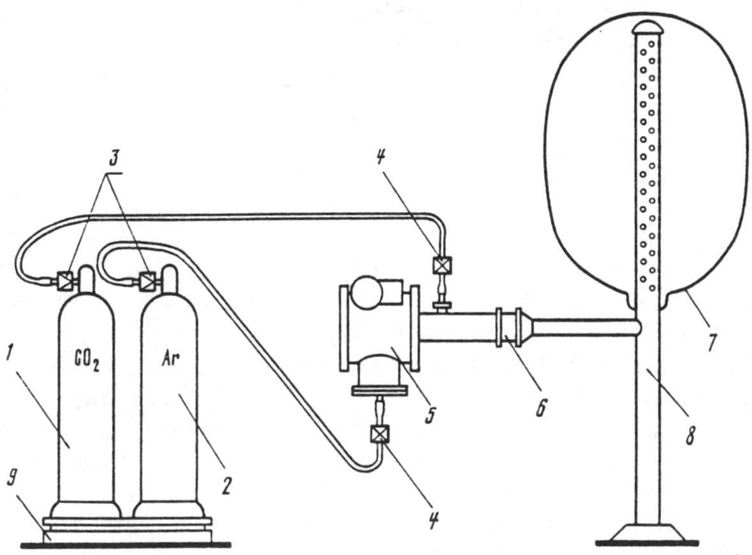

Fig. 6. Diagram of the gas mixture inlet system.
1, 2) cylinders, *3*) reducers, *4*) valve, *5*) DU-80 vacuum valve, *6*) gauging collar, *7*) storage balloon, *8*) tube-column, *9*) weights.

In the experiments that were conducted, we used a CO_2:Ar = 1:1 gas mixture, considered optimum in Ref. [10] for lasing at the 03^10-10^10 transition. No preliminary purification of the gases was done. At room temperature the water vapor saturated pressure is 18 Torr. If we assume that there is saturated water vapor in the CO_2 and Ar cylinders, then in relation to CO_2, it is quite valid to take the value 0.02% by volume as an upper estimate for the amount of water vapor in the gas mixture being employed.

1.4 Heater Electrical Supply System

The electrical power system works to produce the required Nichrome tube temperature in the heater and maintain it during the operation of the apparatus. A three-phase voltage-controlled current serves as the power source. Figure 7 shows a diagram of the power supply portion of the system. Switchable semiconductor T-1000 rectifiers are used to rectify the electric current. They are connected in a six-pulse rectification scheme with an artificial null that permits us to avoid current in the neutral conductor of the power supply. The maximum possible pulsed current power imparted to the load by this system is calculated as

$$W = (U_g^2/R)\,(1 + 3\sqrt{3}/2\pi) \approx 1{,}83\,U_g^2/R,$$

where U_g is the effective power supply voltage and R is the load impedance (0.35 ohm). In the experiments conducted on the CM-GL apparatus, a quite low power of 35 kW was used, which corresponds to 5% of the maximum possible value (Fig. 8). The effective current through the load here was 300 A, and the effective current in each phase was 200 A. Measuring the current in phase C was done by an ammeter connected in series through the current transformer. The current through the load was determined by scaling the ammeter readings in a ballast resister R = 22 Ω circuit connected in parallel to the load (see Fig. 7). A special circuit to control the thyristors was designed to synchronize the thyristor turn-on times. The circuit simultaneously produces trigger pulses to the two thyristors making up the circuit, for example VS_{A1} and VS_{C2}, at the instant in time that provides a current pulse through the load in duration $T/12 \approx 1.7$ ms. The thyristors were turned off by a reverse polarity voltage. Figure 8 shows the timing diagram of the thyristor control pulses, and also the current pulse through the load and the voltage on the heater contacts in relation to the phase voltages U_A, U_B, U_C.

Unfortunately, failures occurred in the process of using the heater that were caused by electric breakdown between the Nichrome tubes and the heater housing. An electric arc arose between the tubes and components of the housing, which lead to fusing and destruction of the tubes. It is rather complex to completely eliminate the causes leading to breakdown, since the components are heated to a temperature of 1000 K and are both at atmospheric pressure and in a vacuum of a few Torr. The following measures were taken to minimize the possibility of electric breakdown:

1) ceramic collars were placed at the top of the Nichrome tubes to

increase the discharge gap (see Fig. 5);

2) an electric power supply was selected for the heater in which the voltage on the tubes changes polarity relative to ground after 3.3 ms (see Fig. 8);

3) a fault interrupt circuit was designed for the heater electrical power supply (see Fig. 9).

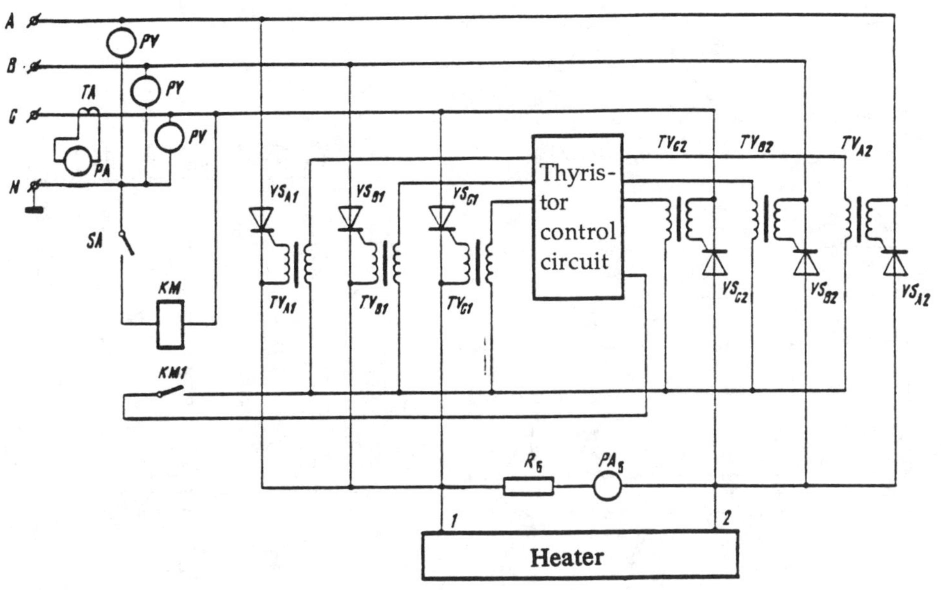

Fig. 7. Diagram of the heater electrical power supply.
VS_{A1}, VS_{B1}, VS_{C1}, VS_{A2}, VS_{B2}, VS_{C2} are T-1000 thyristors; TV_{A1}, TV_{B1}, TV_{C1}, TV_{A2}, TV_{B2}, TV_{C2}, are pulse transformers; *KM* is a magnetic starter; R_b is the ballast resistor (equivalent to the load); PA_b is the ammeter in the ballast resistor circuit; *PA* is the ammeter with the current transformer *TA*; and *PV* are voltmeters.

The power supply's fault interrupt circuit consists of a bridge circuit, in two legs of which two groups of Nichrome tubes are connected along eight pieces, and ballast resistors are connected in the other two legs. During normal operation of the heater, the bridge is balanced and the relay contacts are open. After breakdown of one of the tubes on the housing, the current distribution in the heater changes, the bridge balance is lost, which causes the relays to operate, thus interrupting the delivery of trigger pulses from the control circuit to the thyristors.

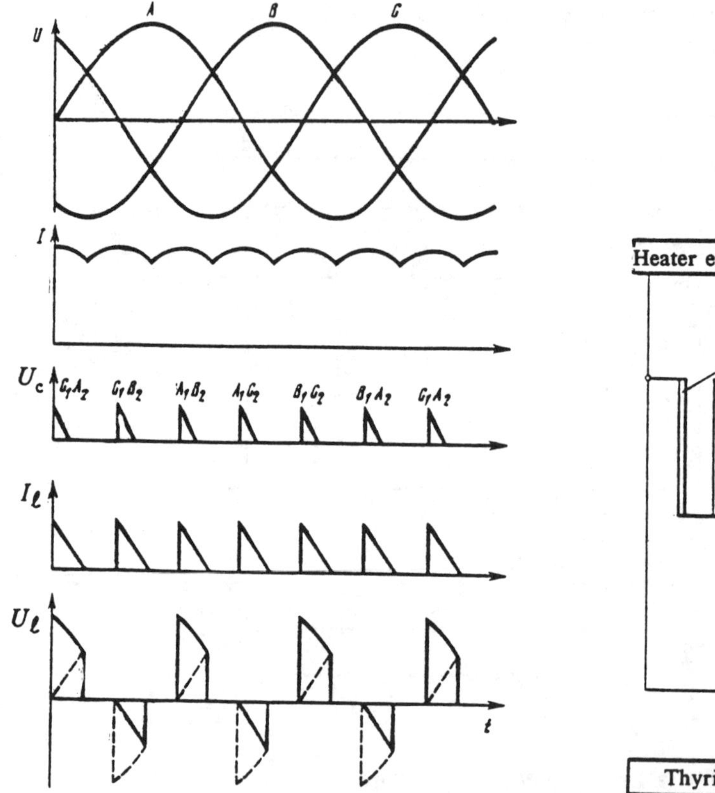

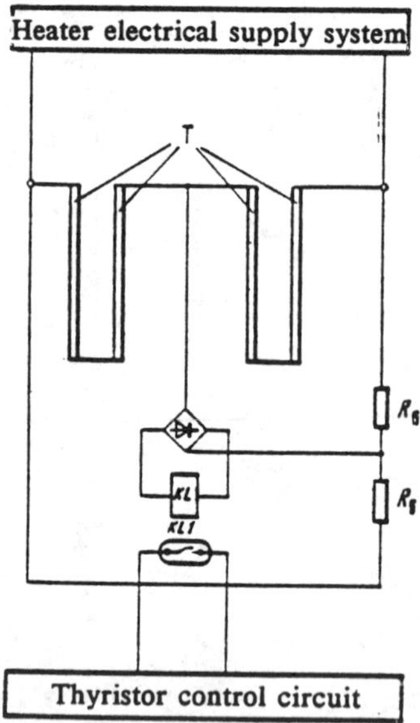

Fig. 8. Timing diagram of the control pulses and the heater's electrical supply current.
U are the phase voltages (U_A, U_B, U_C); I is the maximum possible current through the load; U_c are the trigger pulses that turn on the pairs of thyristors from the control circuit; I_ℓ is the effective current through the load; U_ℓ is the voltage at the contacts *1* (solid curve) and *2* (dashed curve) of the heater relative to ground.

Fig. 9. Fault interrupt circuit for the heater's electrical supply.
T are the Nichrome tubes of the heater, KL is a Herkon relay, R_b is the bridge circuit's ballast resistor.

2. Experimental Results

In the experiments on the CM-GL, lasing was produced in a gas mixture of CO_2:Ar = 1:1 with a stagnation pressure of P_0 = 760 torr (the pressure drop between the reservoir-balloon and the collector was significant due to the large ratio of total cross-sectional area for gas to pass into the heater to the slit area of the nozzle system). The stagnation temperature was varied within 550-700 K. The gas pressure in the resonator was 1.5 Torr. The form of the gas-dynamic flow was observed by the shadow pattern of a HeNe-laser beam illuminating the resonator. Uniform stable flow with trailing lateral shock waves was visible on the shadowgraph. These shock waves separated the zone of the supersonic core from the subsonic portion of the flow. No packing discontinuities in the resonator zone in the supersonic jet were observed. The Mach disk was located outside the resonator; when the instrument was operated for more than 1.5 min, the Mach disk appeared inside the resonator and was observed to move toward the slit.

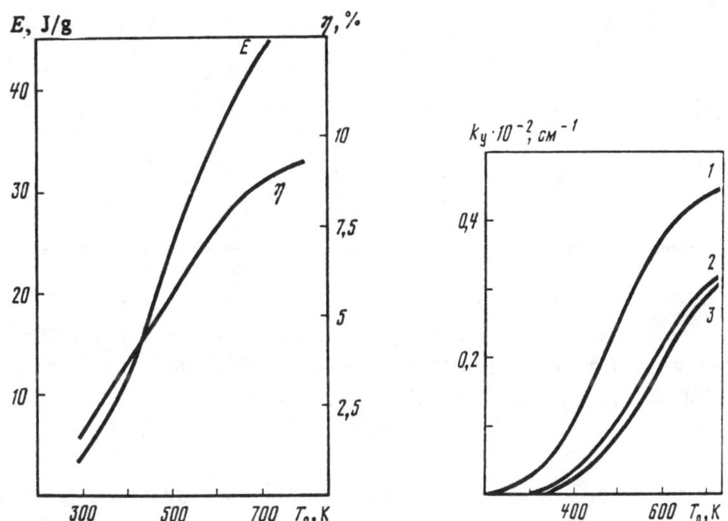

Fig. 10. Variation of the specific output energy E and thermal pumping efficiency η as a function of stagnation temperature for a CO_2:Ar = 1:1 mixture (P_0 = 760 Torr) at a distance of 100 calibers from the nozzle.

Fig. 11. Variation of the gain coefficient for different branches of the 03^10-10^00 transition. CO_2:Ar = 1:1 mixture, P_0 = 760 Torr, 100 caliber distance; *1, 2, 3*) are the $Q(6)$-, $R(6)$-, $P(6)$-branches, respectively.

To evaluate the proposed output parameters, preliminary theoretical calculations were done using the technique described in the first article of the present collection, using the refined values of the relaxation rate constants taken from Ref. [23]. The results of the calculations are shown in Figs. 10-13. All values were calculated for a distance of 100 calibers from the nozzle slit, which coincides with the optical axis of the resonator. Figure 10 shows the variation of the specific output energy and efficiency of thermal pumping as functions of the stagnation temperature. The specific output energy is referenced to the mass of CO_2 in the composition of the gas mixture. The variation of the specific output energy with gas temperature is shown in Fig. 13. The calculated variations of the gain coefficient for *Q*-, *P*-, and *R*- branches of the 03^10-10^00 transition are shown as a function of stagnation temperature in Fig. 11.

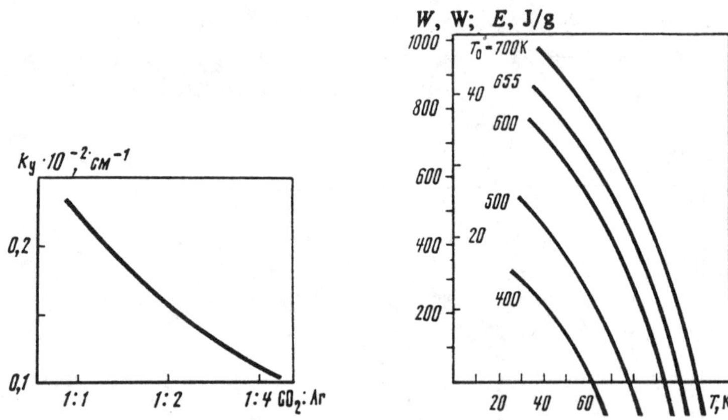

Fig. 12. Variation of the gain coefficient of the *Q*(6)-branch of the 03^10-10^00 transition as a function of mixture composition for T_0 = 500 K and P_0 = 760 Torr at a distance of 100 calibers.

Fig. 13. Calculated variations of the specific output energy and radiant power of the CM-GL apparatus as a function of gas temperature for a CO_2:Ar = 1:1 mixture (P_0 = 760 Torr).

The output power and wavelength were measured in the aforementioned experiments. The wavelength was measured by reflection from a diffraction grating with a grating constant of d = 25 μm. Radiation was observed only at a wavelength of 18.4 μm. The output power was measured by an IMO-2 instrument. Since the IMO-2 stabilizes after 1.5

min in the power measurement regime, the measurements were conducted in the energy measurement regime. It took 10 seconds to start the apparatus; after this time the total radiant energy was measured and then the average output power was calculated. Since the device was not intended for measuring the radiant energy for such a prolonged period of time, to remove systematic errors due to thermal relaxation in the measurement head of the IMO-2, an appropriate calibration of the IMO-2 was performed using a cw electric-discharge CO_2-laser in a similar operating regime and a calibration curve was plotted.

The maximum output power registered during the experiments was 5 W with a stagnation temperature of 655 K. Starting from theoretical calculations without considering the condensation of CO_2 molecules, the gas temperature on the axis of the resonator should be 38 K, and according to Fig. 13 the calculated output power should be 850 W. The disagreement between the theoretical and experimental values can be explained by several causes.

The main cause is the very low efficiency of the resonator. A stable resonator was used in the experiments that was nearly confocal and had a small opening for coupling out the radiation. The choice of the resonator was stipulated by the desire to have a resonator with extremely low losses, however it is not effective for completely converting the energy stored in the gain medium into radiation. The modes of this kind of resonator are situated near the optical axis and occupy a small portion of the entire volume; the fundamental mode occupies less than 0.1% of the resonator and gain medium volume. The field far from the optical axis is small and is not able to convert a significant portion of the population inversion into radiation. This conclusion was verified experimentally when gold-coated glass mirrors were used instead of the copper mirrors. After starting the apparatus and getting it into the operating regime, a portion of the gold coating burned that was roughly 1% of the entire mirror area. The burn on the gold coating indicates that inside the resonator there are areas with a field intensity of no less than 200 W/cm^2 that occupy a small volume.

On more cause that lowers the efficiency of the resonator is the long relaxation time in the coupled mode system. Theoretical calculations show that even under saturated electromagnetic field conditions, complete conversion of the stored energy into radiation occurs over a flow length of more than 10 cm [29].

In addition to the resonator problem, there are also physical causes for the decrease in the energy stored in the gain medium. As was pointed out in Ref. [18], water vapor has a noticeable influence on the rate of the relaxation processes at low temperatures. The effect of condensation is even more substantial according to theoretical calculations [1]. It consists of both the elevated gas temperature due to the liberation of the latent heat of vaporization of CO_2, and the acceleration of the coupled mode relaxation processes. Under conditions realized on the CM-GL apparatus, these causes can lower the energy reserve by 30%, and the calculated output power here is 600 W. If we assume, judging from the gold coating burning through, that the resonator efficiency is roughly 1%, we get acceptable agreement between the calculated and experimental values of the energy stored in the medium.

3. Conclusion

In summary, on the CM-GL apparatus, lasing has been obtained for the first time in the cw regime at the 03^10-10^00 transition of molecular CO_2 at a low stagnation temperature (600-700 K). The low density of the flowing gas in the resonator and the absence of gas-dynamic perturbations demonstrate the possibility of studying in detail the interaction of the gain medium with the optical resonator that make up a GDL.

Note that the CM-GL apparatus permits one to study flows with condensation that are applicable to GDL.

REFERENCES

1. V. K. Konyukhov and V. N. Faizulaev, "Potential for making a gas-dynamic laser based on transitions between levels of the coupled modes of CO_2", KVANT. ELEKTRON., Vol 5 pp 2620-2622, 1978; Preprint FIAN No 204, Moscow, 1978, 11 pp.

2. A. A. Vedeneev, A. Yu. Volkov, A. I. Demin, et al., "Gas-dynamic laser with thermal pumping at transitions between the deformed and symmetric modes of CO_2", PIS'MA ZH. TEKH. FIZ., Vol 4 pp 681-684, 1978.

3. A. A. Likal'ter, "Laser based on transitions between the levels of the coupled modes of CO_2", KVANT. ELEKTRON., Vol 2 pp 2399-2405, 1975.

4. A. A. Likal'ter, "Relaxation of symmetric vibrational modes of CO_2 molecules", ZH. PRIKL. MEKH. TEKH. FIZ., No 3, pp 8-17, 1975.

5. D. G. Bakanov, A. A. Vedeneev, A. Yu. Volkov, et al., "Lasing at the 18.4 μm wavelength in a CO_2 gas-dynamic laser with electric arc heating", KVANT. ELEKTRON., Vol 8 pp 1312-1315, 1981.

6. D. G. Bakanov, A. A. Vedeneev, A. Yu. Volkov, et al., "Gas-dynamic laser with thermal pumping based on transitions between levels of the ν_1 and ν_2 modes of a CO_2 molecule, radiating in the 16.4-17.2 μm region", KVANT. ELEKTRON., Vol 8 pp 1570-1573, 1981.

7. D. G. Bakanov, A. A. Infimovskaya, L. S. Kornienko, et al., "Lasing in the 16.8-17.2 μm wavelength region in a CO_2 gas-dynamic laser", PIS'MA ZH. TEKH. FIZ., Vol 7 pp 802-805, 1981.

8. D. G. Bakanov, A. O. Kulikov, A. I. Odintsov, and A. I. Fedoseev, "Novye lazernye perkhody v molekule CO_2 [New laser transitions in the CO_2 molecule]", in: "Tez. dokl. XI Vsesoyuz. konf. po kogerentnoi i nelineinoi optike [Proceedings of the XI All-Union Conference on Coherent and Nonlinear Optics]", Erevan, 1982, Vol 1, pp 43-44.

9. D. G. Bakanov, A. O. Kulikov, A. I. Odintsov, and A. I. Fedoseev, "Lasing at transitions between upper levels of the symmetric and deformed modes of the CO_2 molecule", PIS'MA ZH. TEKH. FIZ., Vol 9 pp 273-276, 1983.

10. P. Sh. Islamov, Yu. B. Konev, A. O. Kulikov, et al., "Energy characteristics of gas-dynamic lasers operating at transitions between levels of the symmetric and deformed modes of the CO_2 molecule", KVANT. ELEKTRON., Vol 11 pp 551-558, 1984.

11. A. A. Vedeneev, A. Yu. Volkov, A. I. Demin, et al., "Theoretical and experimental investigation of a gas-dynamic laser with thermal pumping based on a mixture of CO_2-Ar(Xe) with an output wavelength of 18.4 $\mu\mu$", Preprint FIAN No 120, Moscow, 1979, 50 pp.

12. M. Brunne, A. Zielinski, J. Milewski, et al., "Analytic modeling of the influence on the nozzle geometry and stagnation conditions of parameters of a cw 18.4 μm gas-dynamic laser", J. PHYS. C, Vol 41 pp 183-187, 1980.

13. M. Brunne, A. Zielinski, J. Milewski, et al., "Influences of the nozzle geometry and stagnation conditions on parameters of a cw CO_2 gas-dynamic laser operating on the $03^10 \rightleftharpoons 10^00$ transition", BULL. ACAD. POL. SCI. SER. SCI. TECHN., Vol 28 pp 653-665, 1980.

14. A. A. Vedeneev, A. Yu. Volkov, A. I. Demin, et al., "Shock-tube driven pulsed CO_2/Ar 18.4 μm gas-dynamic laser", APPL. PHYS. LETT., Vol 38 pp 199-201, 1981.

15. A. A. Vedeneev, A. Yu. Volkov, A. I. Demin, and E. M. Kudryavtsev, "Investigation of an 18.4 μm CO_2-Ne gas-dynamic laser", PIS'MA ZH. TEKH. FIZ., Vol 8 pp 250-255, 1982.

16. D. G. Bakanov, A. I. Odintsov, A. I. Fedoseev, et al., "Laser action on the transition λ = 18.4 μm of the CO_2 molecule in various GDL mixtures", APPL. PHYS. B, Vol 28 pp 288-291, 1982.

17. D. G. Bakanov, L. S. Kornienko, A. I. Odintsov, and A. I. Fedoseev, "Simultaneous quasi-stationary lasing at transitions with wavelengths of 10.6 and 18.4 μm in a CO_2 gas-dynamic laser", ZH. PRIKL. SPEKTROSK., Vol 37 pp 233-236, 1982.

18. A. A. Vedeneev, A. Yu. Volkov, A. I. Demin, and E. M. Kudryavtsev, "Effect of water, hydrogen, and helium diluents on the population of the vibrational levels of carbon dioxide gas under strongly nonequilibrium conditions of supersonic cooling", Preprint FIAN No 26, Moscow, 1981, 24 pp.

19. A. A. Vedeneev, A. Yu. Volkov, A. I. Demin, et al., "Effect of back-pressure on the operation of an 18.4 μm CO_2 gas-dynamic laser", KVANT. ELEKTRON., Vol 9 pp 2332-2336, 1982.

20. A. Yu. Volkov, A. I. Demin, E. M. Kudryavtsev, et al., "Estimation of parameters of a cw 18.4 μm CO_2/Ar gas-dynamic laser", in: "II Intern. Symp. Gas-flow and Chem. Lasers. Sept. 1978", (J. F. Wendt, editor, Hemisphere publ. corp.), 1979, pp 249-252.

21. M. Brunne, A. Zielinski, J. Milewski, et al., "Simplified calculation for prediction of parameters of 18.4 μm CO_2 continuous-wave gas-dynamic lasers", J. APPL. PHYS., Vol 52, pp 74-76, 1981.

22. Yu. B. Konev, "Gas-dynamic laser amplifier with partial inversion at a wavelength of 16 μm", ZH. TEKH. FIZ., Vol 49 pp 1918-1923, 1979.

23. R. Sh. Islamov, Yu. B. Konev, N. I. Lipatov, and P. P. Pashinin, "Theoretical study of the characteristics of a gain medium operating on the transitions between levels of symmetric and deformed modes of CO_2 with thermal pumping", Preprint FIAN No 113, Moscow, 1982, 35 pp.

24. V. K. Konyukhov and V. N. Faizulaev, "Relaxation of the energy

of anharmonic oscillators and coupled modes of the CO_2 molecule", Preprint FIAN No 89, Moscow, 1981, 15 pp.

25. A. A. Likal'ter, "Vibrational kinetics of CO_2 with strong excitation of Fermi-resonant modes", TEPLOFIZ. VYS. TEMP., Vol 20 pp 614-620, 1982.

26. A. Yu. Volkov, A. I. Demin, E. M. Kudryavtsev, and N. N. Sobolev, "Gazodinamicheskie lazery na sparennykh modakh CO_2 [Gas-dynamic lasers operating on CO_2 coupled modes]", in: "Vysokotemperaturnaya gasodynamika, udarnye truby i udarnye volny [High-Temperature Gas-Dynamics, Shock Tubes and Shock Waves]", ITMO AN BSSR, Minsk, 1983, pp 92-101.

27. V. A. Akimov, A. Yu. Volkov, A. I. Demin, et al., "CW gas-dynamic laser operating on a mixture of CO_2-Ar with an output wavelength of 18.4 μm", KVANT. ELEKTRON., Vol 10 pp 886-889, 1983.

28. A. N. Baranov, A. Yu. Volkov, A. I. Demin, E. M. Kudryavtsev, "Electro-gas-dynamic laser operating on transitions between levels of coupled modes of the CO_2 molecule", KVANT. ELEKTRON., Vol 12 pp 2202-2203, 1984.

29. L. A. Fomenko, "Vzaimodeistvie kolebatel'no-neravnovesnogo uglekislogo gaza c rezonansnym izlucheniem na perkhodakh mezhdu urovnyami svyazannykh mod [Interaction of vibrationally-nonequilibrium carbon dioxide gas with resonant radiation at the transitions between levels of coupled modes]", Author's abstract of dissertation for cand. phys.-math. sci., Moscow, 1986, 12 pp.

QUASI-STATIONARY APPROXIMATION IN THE VIBRATIONAL KINETICS OF ANHARMONIC OSCILLATORS

V. N. Faizulaev

The kinetics of the population of the vibrational levels of diatomic molecules allowing for anharmonicity has been examined, beginning with Ref. [1], in many works. In the most typical formulation, the problem reduces to seeking the quasistationary distribution function of anharmonic oscillator (AO) levels in the presence of a hierarchy of *VT*- and *VV*-exchange rates that are spectrally differentiated [2]. The solution is usually done analytically by confining oneself to the zeroth approximation relative to excitation flux (quasi-equilibrium) [1-6]. A classical example of this approach is the Treanor kinetics model [1], in which quasi-equilibrium is assumed to dominate single-quantum *VV*-exchange. In this case, the anharmonicity leads to an unbounded growth in the population $\overline{N}_v$ for levels v above some critical level v_*. It is assumed that the correct (finite) distribution here can be obtained by only postulating the possibility of *VT*-processes and resonant *VV*-exchange being dominant [2-6].

However, there is a more general reason for the instability of the ascending branch of $\overline{N}_v$. It is related to the (nonlocal) action of *VT*-processes causing the transfer of excitation over levels during *VV*-exchange. The quantum flux facilitates the formation of a plateau in the region $v > v_*$ [7]. This effect is in many ways analogous to the flux instability of a quasi-equilibrium cluster size distribution during the condensation of a supersaturated vapor. The theory of its applicability to processes of gradual formation and breakup of clusters was formulated quite some time ago on the basis of both a discrete and a continuous representation [8, 9]. In vibrational kinetics, despite the closed-from description of gradual processes, the latter approach has been worked on the most, which is suitable only for analyzing the regime of strong nonequilibrium with the resonant channel of *VV*-exchange [7, 10]. In Ref. [11], a discrete variant of the flux model of AO relaxation was proposed that is mathematically similar to the approach used in nucleation theory

[8]. This helped to correctly examine the regime of strong nonequilibrium with a substantially nonequilibrium channel of *VV*-exchange, and even to clarify the features of quasi-stationary relaxation of randomly degenerate vibrations of CO_2 molecules, where the anharmonicity is expressed as a strong Fermi-splitting of the energy levels [12, 13]. Remaining unresolved, however, was the problem of explicitly accounting for the attenuation of the quantum flux during *VT*- processes and multi-quantum *VV*-exchange, which play a large role in the population kinetics of highly excited states. The present work is also devoted to the corresponding development of the quasi-stationary approximation on the basis of the flux model [11, 14].

1. Description Technique

We shall formulate the description of AO relaxation kinetics for a single-component gas. We shall start from the equations of balance for the populations of the levels n_v, allowing for the possibility of multi-step transitions $v = v \pm 1, v \pm 2, \ldots, v \pm \Delta$ for *VT*- and *VV*-exchange:

$$dn_v/dt = I_v - I_{v+\Delta}, \quad v = 0, \ldots, D, \tag{1}$$

$$I_v = \sum_{s,p=0}^{\Delta} I^{(p)}_{v-s,v}, \quad I^{(p)}_{v-s,v} = n_{v-s} R^{(p)}_{v-s,v} - n_v R^{(p)}_{v,v-s}, \tag{2}$$

where

$$R^{(p)}_{v-s,v} = \frac{Z}{N} \sum_{m=p}^{D} n_m Q^{m,m-p}_{v-s,v}, \quad R^{(p)}_{v,v-s} = \frac{Z}{N} \sum_{m=0}^{D} n_m Q^{m,m+p}_{v,v-s}$$

are the partial component rates of the transition $R_{vv'} = \sum_{p=0}^{\Delta} R^{(p)}_{vv'}$; $Q^{m\,m'}_{vv'}$ is the probability of the process during which one of the interacting oscillators makes a transition from level m to level m', and the other oscillator goes from v to v'; Z is the frequency of collisions; and $N = \Sigma n_m$ is the total number of molecules. For *VT*-exchange ($p = 0$) it is assumed that the transition probability does not depend on the state of one of the oscillators: $Q^{m\,m'}_{vv'} = P_{vv'}$.

This system of equations can be treated as a finite-difference analog of the continuity equation for the number of molecules, describing the transfer of excitation over levels during *VV*- and *VT*-processes. The quantity I_v characterizes the excitation flux at level v and n_v

characterizes the concentration. Excitation can be conveyed both by molecules and by quanta. A feature of the quantum excitation flux is the fact that it can exist independently of the molecular excitation flux, particularly in the absence of the latter. This is related to the action of a negative source of quanta at higher levels, the role of which is played by the *VT*-process. The basic motive force for the quanta is *VV*-exchange as a process capable of maintaining their number constant. For molecules, *VT*-exchange can also perform this function, while mere dissociation plays the role of a negative source.

Thus, in the framework of the continuity equation, the problem reduces to determining the distribution function formed under the action of negative quantum and molecular sources of excitation. We shall be interested in the (quasi-)stationary case. It can be analyzed best in stages, confining ourselves first to examining the simplest flux model [11]. In this model, in analogy with the Treanor model, it is assumed that only single-quantum transitions take place in AO with *VV*-exchange predominant in the portion of the spectrum up to level K, and with *VT*-exchange predominant in the upper portion of the spectrum from level d ($d > K$), i.e.,

$$R_{v,v+1} = R^{(p)}_{v,v+1}, \quad R_{v+1,v} = R^{(s)}_{v+1,v}, \tag{3}$$

where $p = s = 1$ when $v < K$ and $s = p = 0$ when $v > d$. For the intermediate zone $v = K, K+1, \ldots, d$, where both processes can be effective in which $p = 1$ and $s = 0$, respectively, in Eq. (3). The latter condition permits us to treat the level $K+1$ as a quantum "trap", ensuring that they are conducted completely out of the system due to *VT*-exchange.

We shall assume that the level $D+1$ plays a similar role for molecules as a consequence of dissociation. In these approximations, the description of the stationary kinetics reduces to conservation equations for the quantum (*VV*) and molecular (*VT*) fluxes individually in the corresponding spectral regions:

$$I^{(p)}_{v,v+1} = I_f, \quad v = g, g+1, \ldots, f, \tag{4}$$

where $p = 1$ for $g = 0$, $f = K$ and $p = 0$ for $g = d$, $f = D$.

It is usually assumed that the molecular flux is small in comparison with the quantum flux. Then the solution to Eq. (4) can be done independently for each portion of the spectrum, which when using the null boundary condition ($n_{f+1} = 0$) gives [11]

$$n_\upsilon/N_\upsilon = 1 - \sigma_{gf}/\sigma_{g,\upsilon-1}, \quad \upsilon = g+1, \ldots, f, \tag{5}$$

$$N_\upsilon = n_g \prod_{m=g+1}^{\upsilon} \frac{R_{m-1,m}}{R_{m,m-1}}, \quad \sigma^{-1}_{\upsilon\upsilon'} = \sum_{m=\upsilon}^{\upsilon'} \frac{N_\upsilon R_{\upsilon,\upsilon+1}}{N_m R_{m,m+1}},$$

$$I_f = n_\upsilon R_{\upsilon,\upsilon+1}\sigma_{\upsilon f}. \tag{6}$$

Note that this solution, by virtue of the nonlinearity of the kinetics equation, has an iterative calculation procedure with the flux distribution normalized to the number of quanta L in the system:

$$\frac{1}{N} \sum_{m=0}^{D} m n_m = L.$$

In addition, as a consequence of the stationarity of the problem, it is assumed that the disappearance of molecules and quanta from the trap levels $f+1$ is compensated by their appearance at level $g+1$. In the absence of external excitation sources, the stationarity condition can be formulated as before, requiring that the relaxation time of the number of molecules and quanta be much greater than the characteristic time τ_f to establish the flux distribution. This time is determined for gradual processes as [15]

$$\tau_f = \sum_{m=g+1}^{f} \frac{n_m}{I_f} \frac{\sigma_{gf}}{\sigma_{g,m-1}}. \tag{7}$$

Given above was a description of the flux model with a finite "trap" of quanta in the region K - d, which permitted us avoid directly examining the effect of the attenuation of the quantum flux. In order to calculate this, it is necessary to make use of the generalized continuity equation for the number of quanta in the oscillator-thermostat system

$$I_\upsilon = I^{(1)}_{\upsilon,\upsilon+1} + I^{(0)}_{D,D+1}, \quad \upsilon = 1, \ldots, D, \tag{8}$$

that allows for the action of negative VT-sources of excitation distributed over the spectrum. Solution (8) can be found by the ADI method, as is done for tridiagonal matrix equations [16]. In this case, if we neglect the molecular flux in Eq. (8), we obtain

$$n_\upsilon = n_0 \prod_{m=1}^{\upsilon} x_m, \tag{9}$$

$$x_{m-1} = \frac{R_{m-1,m}}{R_{m,m-1} + j_{m,m+1}}, \quad j_{m,m+1} = R^{(1)}_{m,m+1} - x_m R^{(1)}_{m+1,m}.$$

Equation (9) gives a double recursion calculation procedure in which all levels are first run downward and the values of the quantities x_m and the individual fluxes $j_{m,m+1}$ are found, and then the desired values of n_v are determined in the reverse direction. The boundary condition here in the region $v > K$ can be assigned in the flux "trap" approximation, i.e., assuming $j_{v,v+1} = 0$, which corresponds to the assumption of complete closure of the quantum flux on the thermostat due to *VT*-exchange at the boundary transition. It can be shown that for the simplest flux model the solution (9) is identical to Eqs. (5) and (6).

This calculation scheme can be generalized to the case of several *VV*-exchange channels, when there is no extended interaction over the spectrum between them that is caused by the dissociation of quanta in the system. Indeed, this situation is typical for AO, where single-quantum transfer of excitation (predominant in the lower portion of the spectrum) usually changes sharply into two-quantum exchange in the upper portion. In this case, the stationary kinetics equation

$$I_v = \sum_{s,p=1}^{\Delta} I^{(p)}_{v,v+s} \tag{10}$$

can be transformed into tridiagonal matrix form with solution (9), in which it is useful to set

$$R^{(1)}_{v-1,v} = \sum_{s,p=1}^{\Delta} \frac{x_{v-s}}{x_{v-1}} R^{(p)}_{v-s,v}, \quad R^{(1)}_{v,v-1} = \sum_{s,p=1}^{\Delta} R^{(p)}_{v,v-s}. \tag{11}$$

It is assumed that the iterational solution is sought first in the single-quantum transition approximation, and then, considering the effective rates of the gradual processes determined above, the desired flux distribution during multi-quantum exchange is calculated from Eq. (9).

2. Transition Probabilities

In the vibrational kinetics of AO, the transition probabilities are taken for levels of energy E_v of a Morse oscillator

$$E_v = vE_1 - \Delta E v(v-1), \tag{12}$$

where $\Delta E = E_1/2D$ is the anharmonicity constant and $D = 2E_D/E_1$ is the dissociation level. Frequently, however, following the known Landau-Teller model or its analogues [2], the calculation of the perturbation matrix elements $M_{vv'}$ is done in terms of harmonic oscillator (HO) wave functions with the expansion of the interaction potential $V(r)$ in a power series in αy, where α^{-1} is the effective radius of the potential and y is the amplitude of the oscillations. It is assumed that if $\alpha y \ll 1$, then it is possible to restrict ourselves to the contribution of only the sth term of the series for $V_{v,v+\sigma}$:

$$V_{v,v+s} \sim \frac{\alpha^s}{s!}\left(\frac{E_{v+1}}{2M\omega^2}\right)^{s/2}, \tag{13}$$

and to approximately account for the anharmonicity by introducing the variation of the frequency of the oscillations $\omega = (E_{v+1} - E_v)/\hbar$ and energy E_v with number of the level in accordance with Eq. (12).

This approach imposes a strict requirement on the level of excitation of the oscillations. In addition, the condition that αy be small in AO in general does not make it unnecessary to account for the contribution to Eq. (13) of all terms of the power series. An example of this is the short-range interaction of homonuclear molecules with an exponential repulsive potential in which (due to the interference of terms) multi-quantum transitions during collinear collisions are prohibited [17].

We shall examine the effect of anharmonicity on the vibrational transition probability factor in more detail. We shall perform the calculation for homonuclear molecules modeled as Morse oscillators. We shall take the short-range part of the repulsive potential of the type

$$V(r) = V_0 \exp(-\alpha r) \tag{14}$$

to be responsible for the vibrational transition, with an approximation to r that is linear in y that allows for the relative orientation of the molecules in the two-dimensional case:

$$r = r_0 - \Sigma\lambda_i y_i, \quad \lambda_i = ½|\cos\theta_i|, \tag{15}$$

where r, r_0 are the distances between the nearest atoms of the colliding molecules and the centers of mass, respectively, and θ_i is the angle between the axis of the ith molecule and the radius vector $\mathbf{r}_0$. This enables us to represent the transition probability in the following form [17]:

$$Q_{vv'}^{mm'} = Z_{vv'} Z_{mm'} \bar{Z}, \tag{16}$$

$$Z_{vv'} = \langle |V_{vv'}|^2 \rangle, \quad V_{vv'} = \int_{-\infty}^{\infty} \psi_v(y) \exp(\lambda \alpha y) \psi_{v'}(y)\, dy,$$

$$\bar{Z} = \left\langle \left| \frac{V_0}{\hbar} \int_{-\infty}^{\infty} \exp\left[-\alpha r_0(t) + i \frac{|\Delta\epsilon| t}{\hbar} \right] dt \right|^2 \right\rangle,$$

$$\Delta\epsilon = E_m + E_v - E_{m'} - E_{v'}.$$

The averaging in the vibrational factor $Z_{vv'}$ is done over relative orientations, and the translational averaging $\bar{Z}$ is done over molecular velocities. In the basis of wave functions ψ_v of the Morse oscillator, the matrix elements of the perturbation can be calculated analytically for any transition. The choice of this interaction potential is also tied mainly to this fact. The general expression for $V_{v,v+s}$ has the form [18]

$$V_{v,v+s} = 2(-1)^{v+s} \left[\frac{(D-v)(D-v-s)}{v!(v+s)!} (2D-v-s)_s \right]^{1/2} C_s(\kappa),$$

$$C_s(\kappa) = \sum_{n=0}^{v} (-v)_n \frac{\Gamma(2D-2v-s-1+\kappa+n)}{n!\,\Gamma(2D-2v+n)} (-v+\kappa+n)_{v+s}, \tag{17}$$

$$\kappa = -2\lambda\alpha/\alpha',$$

where $(v)_n = v(v+1) \ldots (v+n-1)$ is the Pochhammer symbol,and α' is a parameter characterizing the repulsive part of the intermolecular Morse potential $V' \sim \exp(-\alpha' y) - 2\exp(\frac{1}{2}\alpha' y)$. For homonuclear molecules we shall assume that $\alpha = \alpha'$ [17].

Equation (17) can be simplified by using, as in Ref. [18], the asymptotic expansion of the ratio of the gamma functions $\Gamma(z+a)/\Gamma(z+b)$ for $z \gg 1$ [19]. Then, assuming $z = 2D - 2v$, accurate to terms of order n^2/z^2, we get

$$C_s(\kappa) = (-1)^v (s+1)_v (\kappa)_s z^{-\bar{s}} a, \tag{18}$$

$$a \simeq 1 + \frac{\bar{s}(\bar{s}+1)}{2z} - \frac{(\kappa+s)v\bar{s}}{(s+1)z} + \frac{(\kappa+s)(\kappa+s+1)\bar{s}(\bar{s}+1)v(v-1)}{(s+1)(s+2)\,2z^2},$$

$$\bar{s} = s + 1 - \kappa.$$

This calculation, judging from the accuracy given for calculating C_s by Eq. (17) (see Table 1), is valid if $v < z/s$. For $\kappa = -1$, when there is a prohibition on multi-quantum transitions, it is precise for all levels.

Table 1

Precise values of the correction a calculated from to Eq. (17) for $\kappa = 0.5$ and $D = 30$

s	v/D				s	v/D			
	1/3	1/2	2/3	3/4		1/3	1/2	2/3	3/4
1	0.79	0.64	0.47	0.35	3	0.58	0.34	0.15	0.066
2	0.67	0.46	0.26	0.15	4	0.52	0.25	0.086	0.029

For homonuclear molecules, this case corresponds to the Landau-Teller model of collinear collisions and is described by the vibrational factor

$$Z_{v,v+s} = A_{v,v+s} Z^0_{v,v+s}, \tag{19}$$

where

$$Z^0_{v,v+s} = (2D)^{-s}(v+1)_s/(s!)^2;$$
$$A_{v,v+1} = D^3(D - v/2)/(D-v)^4$$

are terms differing by zero and $Z^0_{vv'}$ is the value of $Z_{vv'}$ in the harmonic approximation. In general, as a consequence of the anisotropy of the potential (14), all transitions should be resolved as in the harmonic oscillator. Considering Eq. (18) here we have

$$A_{v,v+s} = \bar{a}^2 \frac{D^s(D - v/2)^s}{(D-v)^{2s}} B_{v,v+s}, \tag{20}$$

$$B_{v,v+s} = 4D^2 \int_0^{\pi/2} (\kappa)^2_s z^{2\kappa} \sin\theta \, d\theta = 4D^2 \langle (\kappa)^2_s \rangle,$$

where

$$\langle \kappa^n \rangle = (2 \ln z)^{-n-1} \gamma(n+1, 2\ln z);$$

$\bar{a}$ is the value of a for $\bar{\kappa}$, corresponding to the maximum of the function inside the integral sign, and $\gamma(n, x)$ is the incomplete gamma function [19].

The factor $B_{vv'}$ here acts as a spatial probability factor characterizing the collision efficiency averaged over molecular orientations. In an anharmonic oscillator, collisions with $\overline{\kappa} \simeq (\ln z)^{-1}$ are the most favorable for vibrational transitions, i.e., those collisions at an angle close to $\pi/2$ for $z >> 1$ and not close to π as in a harmonic oscillator. This also causes a possible difference between $Z_{vv'}$ and $Z^0_{vv'}$ at lower levels. When u is increased to the spatial factor, the amplitude effect of the anharmonicity is increased, which usually is calculated from Eqs. (12) and (13). Here it is described by the factor with $B_{vv'}$ coinciding with that calculation only for low excitation levels, when $a \simeq 1$ (see Table 1).

As for the effect of the anharmonicity on $\overline{Z}$, it is determined primarily by the nonequivalence of the Morse oscillator spectrum (12) and in the adiabatic approximation as a function of the energy defect $\Delta\epsilon$, or more precisely, as a function of the parameter

$$\gamma = \frac{\pi |\Delta\epsilon|}{\hbar\alpha} \sqrt{\frac{\mu}{2T}} \qquad (21)$$

is described in the following manner [17]:

$$\overline{Z} = f(\gamma) \exp\left(\frac{\Delta\epsilon - |\Delta\epsilon|}{2T}\right), \qquad (22)$$

$$f(\gamma) \sim \exp(-{}^2/_3\gamma) \text{ при } \gamma < 20, \quad f(\gamma) \sim \gamma^{7/3}\exp(-3\gamma^{2/3}) \text{ при } \gamma > 20,$$

where μ is the reduced mass of the molecules and T is the gas temperature. From here, in particular, it follows that the probability for the transition $m+p$, $v \rightarrow m$, $v+s$ in the case of VT-processes (p=0) should monotonically increase with increasing v, and for VV-exchange (s, $p \neq 0$) should pass through a sharp maximum in the vicinity of resonance at $v_0 = D(1 - p/s) + (p/s)m$.

This is illustrated in Fig. 1, which shows the probability distribution of the competing processes over the spectrum for nitrogen molecules ($E_1 = 3358$ K, $E_D = 113{,}150$ K, $\alpha^{-1} = 0.2$ Å). The calculation is done using the formula

$$Q^{m+p,m}_{v,v+s} = \overline{A}_{m+p,m}\overline{A}_{v,v+s}(m+1)_p(v+1)_s Q^{p0}_{0s} f(m, v) \qquad (23)$$

using the recurrence relation for the adiabatic function

$$f(m', v') = f(m, v) \exp\{-b[p(m'-m) - s(v'-v)] \operatorname{sgn}(v_0 - v)\},$$
$$b = 2\Delta E \, |d \ln \overline{Z}/d\Delta\epsilon|. \qquad (24)$$

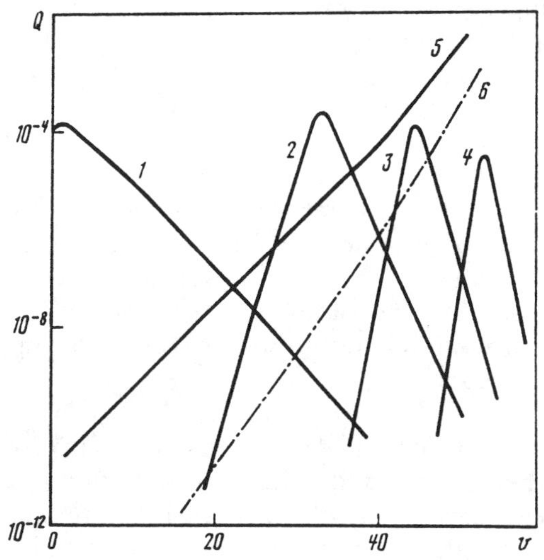

Fig. 1. Probability dependence on the level number for nitrogen molecules at T = 300 K. *1)* $Q^{10}_{v,v+1}$, *2)* $Q^{10}_{v,v+2}$, *3)* $Q^{10}_{v,v+3}$, *4)* $Q^{10}_{v,v+4}$, *5)* $P_{v+1,v}$, *6)* $P_{v,v+1}$

The correction to the anharmonicity $\overline{A}_{v,v+s} = A_{v,v+s}/A_{0s}$ in Eq. (23) is calculated from Eq. (20) using the precise values of $\overline{a}$. Calculated values for lower levels [20] served as references for single-quantum processes, and the probabilities of the resonant channels of VV-exchange correlated with Q^{01}_{10} served as the references for multi-quantum processes.

3. Quasi-Stationary Population Distribution

We shall begin the analysis with the quasi-equilibrium solution to Eq. (4) for single-quantum VV-exchange. Corresponding to this is the Treanor population distribution

$$N_v = N_0 \exp\left[-vE_1/T_1 - (E_v - vE_1)/T\right], \quad v = 0, \ldots, K, \tag{25}$$

in which all components of the excitation flux $I^{(1)}_{v,v+1}$ vanish for

$$Q^{m+1,m}_{v,v+1} = Q^{m,m+1}_{v+1,v} \exp\left[(E_{m+1} - E_m + E_v - E_{v+1})/T\right], \tag{26}$$

where T_1 is the vibrational temperature determined from the condition of (25) for normalizing with respect to the number of quanta. Under nonequilibrium conditions ($T_1 > T$), this distribution for an AO has a minimum at the point

$$v_* = 1/2 + DT/T_1. \tag{27}$$

The existence of this minimum indicates that the quasi-equilibrium is thermodynamically unstable during *VV*-exchange. Actually, from the relation for the rates of the forward and reverse processes, corresponding to Eqs. (25) and (26)

$$R_{v,v+1}/R_{v+1,v} = \exp\left[2\Delta E(v - v_*)/T\right], \tag{28}$$

it follows that it is sufficient for quanta to reach the level v_*, as this should unavoidably lead (for an open *VT*-channel) to the excitation being carried away through the upper levels. In this sense, level v_* is critical, fulfilling in vibrational relaxation the same role (narrow region) that nucleation centers of critical size in the liquid phase fulfill during the condensation of a supersaturated vapor.

Flux effect. It is known from nucleation theory that the flux of particles can strongly distort the quasi-equilibrium distribution of clusters, confining it to the plateau section of the supercritical region. We shall examine this effect in more detail in the case of the vibrational relaxation of an AO during single-quantum *VV*-exchange. As a starting point, we shall cite the solution for the simplest model Eqs. (5) and (6). From the definition of the current intensity in Eq. (6), it follows that the quantity $\sigma_{vv'}$ characterizes the permeability of the chain of transitions v ... v' in relation to the internal excitation flux $n_v R_{v,v+1}$ that cycles at the transition v, v+1. In the Treanor model, this is assumed to equal zero, which corresponds to the condition of total (mirror) reflection of the quanta at the K+1 boundary. In the flux model, the level K+1, conversely, is assumed to be completely absorbing to quanta, as a consequence of which an excitation flux also arises in the transition chain. Its magnitude is greater the shorter the chain is and the higher the permeability of the individual links are. The latter depends on the probability characteristics of the transitions. Here and below it is determined in analytical calculations according to Eqs. (23) and (24) for $\overline{A}_{mm'} = 1$ and for a value of b that is independent of the level number.

The most distinct flux effect appears in the weak nonequilibrium regime, when everywhere in region 1, ... , K long-range *VV* exchange dominates with the rates

$$R^{(1)}_{v,v+1} = (Z/N)n_1 Q^{10}_{v,v+1}, \quad R^{(1)}_{v+1,v} = (Z/N)n_0 Q^{01}_{v+1,v}. \tag{29}$$

It is noteworthy that in this case Eq. (28) is satisfied and, consequently, the quasi-equilibrium distribution N_v should coincide with the Treanor distribution N_v. Taking this into consideration and using Eqs. (5) and (6) we get

$$\sigma_{vv'} = (N_u R_{u,u+1}/N_v R_{v,v+1}) F^{-1}_{vv'}, \quad v \leqslant u \leqslant v', \tag{30}$$

$$F_{vv'} = \frac{1}{2}\sqrt{\frac{\pi}{\lambda}}\exp\left(\frac{b^2}{4\lambda}\right)\{\mathrm{erf}\,[\sqrt{\lambda}(v'-\tilde{u})] - \mathrm{erf}\,[\sqrt{\lambda}(v-\tilde{u})]\},$$

$$\tilde{u} = u + b/2\lambda,$$

where $\lambda = \Delta E/T$, u is the level corresponding to the minimum of the internal flux distribution $N_u R_{u,u+1}$ in the chain v ... v'. For a Treanor distribution, the narrowest place in the entire chain 1, ..., K is $u = v'$. Therefore, for levels $v < v_*$, the flux effect turns out to be small according to Eqs. (5) and (30), whereas the distribution of n_v is close to being a Treanor distribution formed as a result of detailed balance of the opposing processes $v \rightleftarrows v + 1$. In the $v > v_*$ region, the situation is different. Due to the high permeability of the transitions σ_{vK}, inherent here for further VV-exchange, the quasi-stationary distribution in this case is formed predominantly in the processes of gradual VV-activation of the levels $v{-}1 \to v \to v{+}1$. Since the rates of further VV-exchange drop with increasing v (due to the increase in the resonance defect), then despite the significant deviation from being Treanor, the distribution of n_v continues to rise in correspondence with Eq. (6).

With an increase in the population of the levels in the $v > v_*$ region, the role of near-range (resonant) VV-exchange becomes more noticeable and occurs at the rates

$$R^{(1)}_{v,v+1} = (Z/N) n_{v+1} Q^{v+1,v}_{v,v+1}, \quad R^{(1)}_{v+1,v} = (Z/N) n_v Q^{v,v+1}_{v+1,v}. \tag{31}$$

For a sufficient level of excitation for that portion of levels $v > v^*$, this processes begins to dominate over further VV-exchange, imposing an additional limitation on the distribution in the plateau segment in the $v > v_*$ region [2-7, 10]. It is interesting that the quasi-equilibrium distribution in this case

$$N_{v+1}/N_v = v/(v+1) \tag{32}$$

is distorted slightly by the VV-flux:

$$\sigma_{vv'} = 1/(v'-v), \quad n_{v+1}/N_{v+1} = \sqrt{(K-v)/(K-v^*)}. \tag{33}$$

In principle, this should also be expected if one considers the diffusion mechanism of the transfer of excitation during resonant processes. However, the role of this effect is also important in this case, since without violating the detailed balance of the processes as a whole, the partial quasi-equilibrium during near-range VV-exchange, as well as the distribution of Eq. (32), would be impossible.

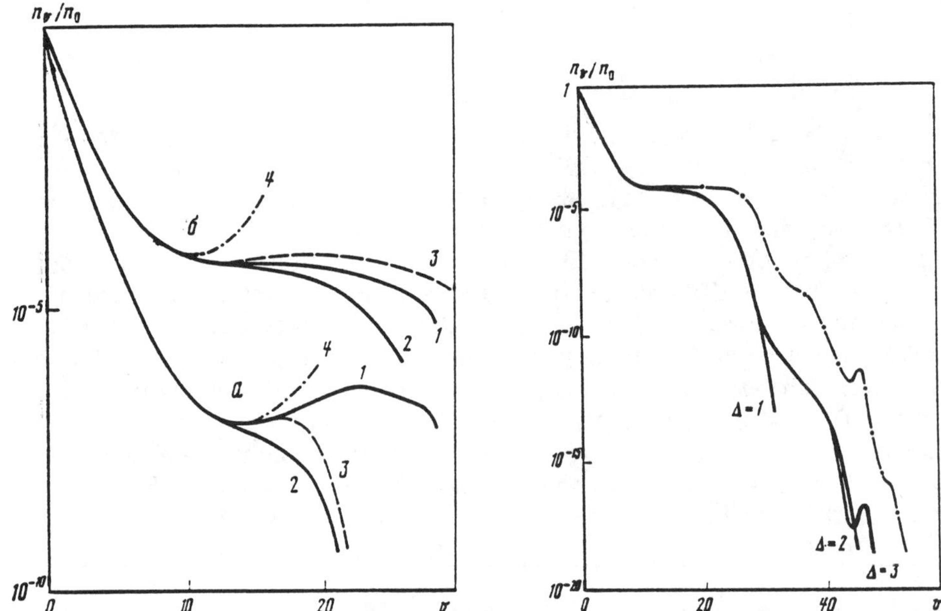

Fig. 2. Population distribution of the levels for single-quantum transitions in N_2 under different conditions.
a) T = 300 K, T_1 = 1500 K, *b*) T = 300 K, T_1 = 2000 K. *1*) flux solution to Eq. (5) for K = 27; *2*, *3*) flux (9) and quasi-equilibrium solution according to the data of Fig. 1, respectively; *4*) Treanor distribution.

Fig. 3. Population distribution of the levels for multi-quantum exchange in N_2 with T = 300 K, T_1 = 2000 K.
Solid curves denote the flux solution (9) in the Δ-quantum approximation and the dot-dashed curve denotes the quasi-equilibrium solution according to the data of Fig. 1.

The (intermediate) regime when $v_* < v^* < K$ is indicative in this sense. Figure 2 (curve *1*, *a*) shows the flux distribution corresponding to this regime for nitrogen molecules. A sharp bend at the point v^* is typical here and is related to the transition from the long-range to near-range VV-exchange and the presence of an ascending branch of n_v in the $v_* < v < v^*$ region, which takes place below the Treanor branch. The later also makes it possible to move the distribution onto a plateau for $v < v^*$. With an increase in the excitation, near-range VV-exchange gradually surrounds the entire supercritical portion of the spectrum $v \geq v_*$ and the flux distribution acquires the form inherent to the strong nonequilibrium regime (see Fig. 2, curves *1, b - 3, b*).

Effect of VT-exchange and multi-quantum VV-exchange. We shall start with *VT*-processes. Their role in population relaxation kinetics is two-fold: on one hand they act as a negative source of excitation in the $v > K$ region, thereby forming (indirectly) the flux distribution at the lower levels, and on the other hand (usually this is only considered in the calculations of Refs. [2-6]) they directly participate in the depopulation (vacating) of the levels ensuring a smooth transition to an equilibrium distribution in the k ... d region. Thus, instead of a finite quantum trap at the K+1 level, a transition zone with strong attenuation of the quantum flux and a drop in the population of the levels appears in the spectrum. The effect of *VT*-processes here can be estimated from the quasi-equilibrium (local) balance of the *VV*- and *VT*-fluxes: $I_v = 0$. Then, according to Eq. (3), taking level K as the lower boundary of the band, where $R^{(1)}_{K,K+1} = R^{(0)}_{K+1,K}$, and d for the upper boundary, where the distribution of n_v is no longer in the equilibrium region, we get

$$n_v = n_K \exp\left[-\lambda(v-K)^2(D-d)/(d-K)\right], \quad v = K, K+1, \ldots, d. \tag{34}$$

In the $v < K$ region, this effect on the population of the levels has a nonlocal character and is related to the transfer of the excitation during *VV*-exchange. Its analysis was given approximately above on the basis of solutions (5) and (6). However, not taken into account in this calculation is the dependence of K on I_K, which was assumed implicitly in the simplest model (in Fig. 2 the value of K is not calculated for curves *1*). Its presence, as calculating the quasi-stationary distribution according to Eq. (9) shows, can in many ways determine the flux effect in the supercritical region (see Fig. 2, curves *2, 3*).

In addition to *VT*-processes, as shown in [3, 21], multi-quantum *VV*-exchange can also play an important role in shaping the population distribution at the upper levels. This is facilitated not only by the high probability of the resonant exchange channels (see Fig. 1), but also by the high kinetic efficiency of multi-stage depopulation of the levels. The later is determined by the factor x_{v-s}/x_{v-1} in Eq. (11), which rapidly increases with increasing s in the decay region of the distribution. Therefore, at high levels in the depopulation kinetics, it is also possible for there to be a predominance of those levels which, for example, as in four-quantum exchange in N_2, are significantly inferior to *VT*-processes in terms of probabilities. The dominance of multi-quantum *VV*-exchange leads to a strong increase in the population of the levels in the upper portion of the spectrum with the formation of distributions in the vicinity of resonances for $v > v_0$, which are similar to the flux distributions in long-range *VV*-exchange at lower levels. The difference consists only in the position of the Treanor minima $v_*^0 = v_0 + v_*/s$ and in the transition band for each of the distribution's clusters. The form of the cascade flux distribution of the populations of the levels in N_2 is shown in Fig. 3.

4. Relaxation and Dissociation Rates

A flux model with distributed excitation sources can also be used to describe the evolution of the quasi-stationary state of oscillators with the dissipation of quanta and molecules in the system:

$$dL/dt = -\sum_{v=0}^{D} I^{(0)}_{v+1,v}, \quad dN/dt = -I_D. \tag{35}$$

Vibrational relaxation. It is known that in a harmonic oscillator (HO), the rate of this process is primarily determined by quantum losses on the lower levels and is characterized in the Landau-Teller equation for L by the constant $G = |d\ln L/dt|$, equal to $G^0 = ZP_{10}$. Anharmonicity leads to the transport of the excitation to the upper levels, and for that reason the value of G can be significantly more than G^0. Quasi-equilibrium approximations are usually used to model this effect [4-6]. How valid this approach is can be judged from the flux distributions of the excitation in N_2 during single-quantum exchange shown in Fig. 4 under different conditions.

Our attention should now be directed to the fact that the local balance of the *VV*- and *VT*-fluxes ($I_v = 0$) in an AO is maintained only at the upper levels. In the central portion of the spectrum, the *VV*-flux turns out to be uncompensated and is established at roughly a level corresponding to that of the simplest model (Eqs. (5) and (6)). For weak nonequilibrium, this flux is usually small in comparison with that which leaves the system from low levels via *VT*-processes. Therefore, the error of the quasi-equilibrium calculation of the relaxation rate, related to the unconsidered transfer of the excitation to the upper levels by *VV*-exchange, is small in this case. For strong nonequilibrium, when the primary quantum losses are localized in the upper levels (see Fig. 4), it is not valid to neglect the flux effect, or more precisely, its effect on the power of the *VT*-sources corresponding to the Le Chatelier reaction. As is evident from Table 2, which cites a calculation of G/G^0 for N_2, this can lead to overstating the relaxation rate by almost an order of magnitude.

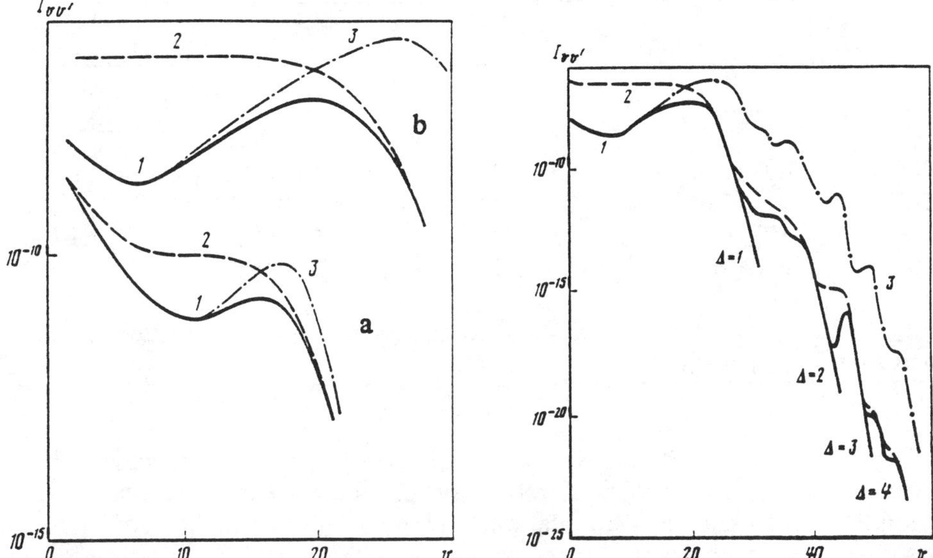

Fig. 4. Distribution of *VT*- and *VV*-fluxes during single-quantum transitions in N_2 under different conditions.
a) $T = 300$ K, $T_1 = 1500$ K; *b*) $T = 300$ K, $T_1 = 2000$ K. *1, 2*) quasi-stationary values of $I^{(0)}_{v+1,v}/NZQ^{10}_{01}$ and $I^{(1)}_{v+1,v}/NZQ^{10}_{01}$ corresponding to solution (9); *3*) quasi-equilibrium distribution of these values.

Fig. 5. Distribution of *VT*- and *VV*-fluxes during multi-quantum exchange in N_2 with $T = 300$ K, $T_1 = 2000$ K.
1, 2) respectively, the quasi-stationary values of $I^{(0)}_{v+1,v}/NZQ^{10}_{01}$ and $I^{(1)}_{v+1,v}/NZQ^{10}_{01}$ obtained for solution (9) in the Δ-quantum approximation; *3*) quasi-equilibrium distribution of these values.

Multi-quantum *VV*-exchange also leads to a drop in the power of the *VT*-sources at the upper levels. However, the effect of this factor is relatively small, to the extent that *VT*-processes lower the suppressing portion of the excitation flux in the region where single-quantum transitions dominate (Fig. 5). The only exception to this is the case of an extremely strong nonequilibrium, when the extent of the plateau region of the first cascade of the population distribution is limited, not by *VT*-processes, but by two-quantum *VV*-exchange.

Table 2

Relaxation and Dissociate Rates of N_2 for T = 300 K

T, K	$\log(G/G^0)$		$\log(I_D^0/I_D)$	
	$\Delta = 1$	$\Delta = 2-4$	$\Delta = 1$	$\Delta = 2-4$
2000	1.0 (2.0)	1.0-1.0	33.2 (21.0)	11.7-4.3
2500	2.4 (3.6)	2.4-2.4	16.1 (9.1)	6.3-2.7
3000	3.0 (4.3)	2.9-2.9	17.4 (5.7)	5.3-2.5
3500	3.2 (4.5)	3.1-3.1	16.9 (5.9)	6.7-3.7

Note: The parentheses denote the quasi-equilibrium values of the corresponding quantities.

Dissociation. When estimating the rate of this process, it is extremely important to allow for both of the factors examined above. An example of a known [2] relation, relating I_D with the equilibrium dissociation rate constant during convective transport of excitation in an AO can be quite convincing. In the model with point molecular sources (Eqs. (5) and (6)), there is a solution analogous to Eq. (30):

$$I_D = n_D R_{D,D+1}, \quad n_D = n_\upsilon \exp\left[-(E_D - E_\upsilon)/T\right] F_{\upsilon D}^{-1}, \tag{36}$$

where $F_{\upsilon D}$ is the correction to the distortion of the equilibrium distribution at the spectral boundary by the molecular flux for $u = D$. The effect of anharmonicity here can be characterized by the quantity I_D^0/I_D, where I_D^0 is the value of I_D with a Boltzmann distribution over all levels with temperature T_1. Calculations show that in the single-quantum approximation the gain to the population of the levels given by VV-exchange in the plateau region is clearly insufficient to compensate for the sharp drop in the distribution in the upper portion of the spectrum caused by VT-processes. As a result, the dissociation rate in an AO turns out to be significantly less than the that in a HO. The (nonlocal) flux effect of VT-processes facilitates this in many ways. Only multi-quantum VV-exchange, forming a cascade flux distribution and thereby narrowing the region where VT-processes dominate at the upper levels (see Figs. 3 and 5), can significantly accelerate dissociation.

Note that, if the calculation is done in the quasi-equilibrium approximation, it is also possible to incorrectly arrive at, as in Ref. 21, for example, the opposite conclusion. The data cited in Table 2 illustrate this. It follows from these data, in particular, that the molecular flux in N_2, when the multi-quantum exchange channels are activated, grows by several orders of magnitude. Because of this, it is possible for the dissociation process to enter a regime close in rate to that calculated for a HO. Two-four-quantum exchange plays the dominant role in this acceleration. In principle, allowing for the high kinetic efficiency of multi-stage excitation transfer, it is necessary to include the substantial influence of other resonant exchange channels as well on the dissociation rate. Accounting for them, however, as well as an entire complex set of factors to which the population of the upper levels is sensitive, requires more thorough analysis of the kinetics and probability characteristics in AO, which is an independent problem.

REFERENCES

1. S. E. Treanor, J. W. Rich, and R. G. Rehm, "Vibrational relaxation of anharmonic oscillators with exchange dominated collisions", J. CHEM. PHYS., Vol 48 No 4 pp 1798-1807, 1968.

2. B. F. Gordiets, A. I. Osipov, and L. A. Shelepin, "Kineticheskie protessy v gazakh i molekulyarnye lazery [Kinetic Processes in Gases and Molecular Lasers]", Moscow, Nauka, 1980, 512 pp.

3. B. F. Gordiets, A. I. Osipov, and L. A. Shelepin, "Kinetic nonresonant vibrational exchange and molecular lasers", ZH. EKSP. TEOR. FIZ., Vol 60 No 1 pp 102-113, 1971.

4. B. F. Gordiets and Sh. S. Mamedov, "Distribution function and relaxation rate of vibrational energy in a system of anharmonic oscillators", ZH. PRIKL. MEKH. TEKH. FIZ., No 3 pp 13-22, 1974.

5. M. B. Zheleznyak and G. V. Naidis, "Distribution of vibrational relaxation and dissociation rates by vibrational levels under nonequilibrium conditions", ZH. PRIKL. MEKH. TEKH. FIZ., No 1 pp 4-9, 1976.

6. Sh. S. Mamedov, "Methods of vibrational kinetics and their application to molecular lasers and laser chemistry", TRUDY FIAN, Vol 107 pp 3-67, 1979.

7. S. A. Brau, "Classical theory of vibrational relaxation of anharmonic oscillators", PHYSICA, Vol 58 No 4 pp 533-553, 1972.

8. R. Becker and W. Doring, "Kinetische Behandlung der Keimbilding in ubersattingten Damphen", ANN. PHYS., Vol 24 p 719, 1935.

9. Ya. B. Zel'dovich, "Theory of formation of a new phase: cavitation", ZH. EKSP. TEOR. FIZ., Vol 12 No 11/12 pp 525-538, 1942.

10. A. A. Likal'ter and G. V. Naidis, "Vibrational distributions in highly-excited molecular gases", KHIMIYA PLAZMY, No 8 pp 156-188, 1981.

11. V. K. Konyukhov and V. N. Faizulaev, "Effect of flux of quanta on the vibrational distribution and relaxation of the energy of anharmonic oscillators", KRAT. SOOBSHCH. FIZ., No 8 pp 36-40, 1981.

12. V. K. Konyukhov and V. N. Faizulaev, "Relaxation of the energy of anharmonic oscillators and coupled modes of molecular CO_2", Preprint FIAN No 89, Moscow, 1981, 20 pp.

13. V. N. Faizulaev, "Effect of flux of quanta on the relaxation of coupled modes of molecular CO_2", ZH. PRIKL. MEKH. TEKH. FIZ., No 6 pp 9-14, 1982.

14. V. N. Faizulaev, "Flux model of vibrational relaxation of diatomic molecules", Preprint IOFAN No 241, Moscow, 1984, 25 pp.

15. R. P. Andres and M. Boudart, "Time lag in multistate kinetics: nucleation", J. CHEM. PHYS., Vol 42 No 6 pp 2057-2064, 1965.

16. D. E. Potter, "Computational Physics", New York, Wiley, 1975.

17. E. E. Nikitin, "Teoriya elementarnykh atomno-molekulyarnykh protsessov v gazakh [Theory of Elementary Atomic-Molecular Processes in Gases]", Moscow, Khimiya, 1970, 456 pp.

18. I. I. Tugov, "Perturbation theory for multi-quantum transitions in diatomic molecules", TRUDY FIAN, Vol 146, pp 17-65, 1984.

19. F. Davis, "The Gamma-Function and Related Functions", (M. Abramovits and I. Stigan, editors), Moscow, Nauka, 1979, pp 80-118, 1979.

20. A. V. Bogdanov, Yu. E. Gorbachev, and V. A. Pavlov, "Theory of the exchange of vibrational and rotational quanta during molecular collisions", Preprint FTI No 833, Leningrad, 1983, 61 pp.

21. A. N. Oraevskii, A. F. Suchkov, and Yu. N. Shebeko, "Effect of two-quantum exchange on the rate of dissociation of a nonequilibrium diatomic gas", KRAT. SOOBSHCH. FIZ., No 1 pp 32-36, 1978.

Part II

ROTATIONAL AND SPIN NONEQUILIBRIUM PROCESSES IN GASEOUS MEDIA

ROTATIONAL NONEQUILIBRIUM PROCESSES IN WATER AND HEAVY WATER VAPOR UNDER SUPERSONIC FLOW CONDITIONS

V. I. Tikhonov

0. Introduction

The first research on rotational molecular relaxation began fairly recently when studying transport phenomena in gases. It turned out to be necessary to allow for rotational relaxation when calculating the transport coefficient of polyatomic molecules. It therefore became necessary to experimentally determine the rotational relaxation time during molecular collisions. Using the temperature transpiration method, suggested by E. A. Mason [1], the rotational relaxation times were measured for N_2, O_2, CO, and CO_2 [2].

The traditional method for determining a relaxation time is measuring the dispersion of ultrasound in molecular gases. Advances at high frequencies permitted the application of the ultrasonic method for studying rotational relaxation [3, 4].

Both of these methods permit the determination of the relaxation time of the mean rotational energy τ_{RT}. More precise research into the rotational relaxation process was made possible with the advent of super-high resolution laser and submillimeter spectroscopy. Methods that are successfully used here are those for studying collisional relaxation processes using optical analogues of nonstationary phenomena, which were earlier observed in nuclear magnetic resonance experiments [5-9].

Of particular interest are the works on four-level double microwave resonance performed by T. Oka and coworkers [10]. They systematically studied a large number of pure gases and gas mixtures and established a series of relationships for rotational transitions during collisions, including collisional selection rules.

With the evolution of experimental gas dynamics, new approaches have appeared for investigating relaxing gases. Supersonic flows turned out to be a suitable object for investigating relaxation phenomena in molecular gases.

First of all, during the expansion of the gas, the temperature in the supersonic flow can attain very low values, which permits one to perform the experiments in a very wide temperature range. Conditions can be realized in the flow which are unattainable in equilibrium systems. For example, as a consequence of the rapid change in the temperature, the gas is not able to condense and there are regions in the flow in which the gas is found in a state of strong supersaturation. A motionless gas under such conditions would condense very quickly, and it would be difficult to study due to the low saturated vapor density.

Secondly, the rate of change of the thermodynamic quantities leads to the violation of equilibrium with respect to the internal degrees of freedom of the molecules that are moving in the flow. Studying the resultant nonequilibrium distribution permits one to draw a number of conclusions regarding the behavior of the relaxation. A gas in supersonic jets of diatomic molecules [11, 12] and polyatomic polar linear molecules [13] was studied by this method.

This work is devoted to studying rotational relaxation of H_2O molecules and D_2O molecules in the gas phase under supersonic jet conditions. This problem arose when studying the potential for producing a water vapor GDL. The authors of Ref. [14] advanced the idea of making a laser that used the energy of condensation of molecules with the hydrogen bond. The essence of the idea consists in that a large number of vibrationally excited water molecule associates (dimers, trimers, etc.) are formed in the initial stage of condensation. It was shown in Ref. [14] that during condensation under supersonic gas flow conditions, a population inversion can appear in the system of vibrational levels of water dimers. Direct observation of the water dimer spectrum is difficult, therefore, an attempt was made to observe the existence of excited dimers by transferring their vibrational energy to the rotational degrees of freedom of H_2O molecules, which should lead to the appearance of rotational nonequilibrium.

Studying supersonic water vapor flowstreams under different initial conditions showed that a violation of the equilibrium distribution is actually observed in the rotational spectrum of H_2O molecules [15]. This nonequilibrium exists both for low initial temperatures, at which condensation occurs at the observation point, and for high temperatures, where condensation is not possible. The most probable cause producing the deviation from equilibrium in this region is the low rate of the relaxation processes for those comparatively low densities at which the studies were done.

Thus, in a supersonic jet for the range of initial conditions being studied, it is necessary to allow for rotational relaxation of water molecules. However, H_2O and D_2O molecules, from the point of view of relaxation phenomena, belong to the most studied type of molecules, namely those with asymmetric rotators. The theoretical works in which collisional relaxation has been described for H_2O and D_2O molecules are extremely few in number [16-18]. Experimental data for these molecules relative to relaxation times in the low temperature range (50-60 K) are entirely lacking.

In Refs. [17, 18], possible selection rules were theoretically predicted for the collisions of H_2O and D_2O molecules. No one has yet confirmed these rules experimentally.

Determining the quantitative characteristics of collisional relaxation processes also permits one to research the possibility of making a GDL based on rotational transitions. These transitions for H_2O and D_2O lie in the submillimeter spectral region, which has few intense sources of laser radiation.

Describing rotational relaxation is also important for understanding the mechanisms of the operation of the cosmic maser, which up to now does not have a generally accepted explanation [19, 20]. And finally, allowing for rotational nonequilibrium of water vapor is necessary when studying processes in the upper, rarefied layers of the atmosphere.

An important point of experimental research is choosing the diagnostics of the rotational level's populations. In Ref. [21], the method of adsorption submillimeter spectroscopy, based on a backward-wave tube (BWT-spectroscopy), was suggested for studying supersonic flows. The advantage to this method consists in that the intensity of the probe ray can be made so low that it does not convey any perturbation to the gas flow. Another advantage is the BWT-radiation frequency can be tuned continuously and has high spectral resolution.

Rotational relaxation is best studied in a gas of the heavy isotope of water, namely D_2O molecules. H_2O and D_2O molecules have identical molecular symmetry, nearly the same multipole moment magnitude, and almost identical rotational spectra, however, the energy levels of D_2O are situated roughly twice as frequently, which permits studying a larger number of transitions using BWT-spectroscopy. In addition, for rotational lines of D_2O, there is almost no absorption in atmospheric air.

It is appropriate to investigate rotational relaxation of H_2O and D_2O molecules in a supersonic flowstream under rapid expansion in two stages: without the condensation process and with condensation. In this work, most of our attention is allotted to the first stage.

1. Description of the V-2 Programmed Control Apparatus and Methods for Determining Gas-Dynamic Parameters in Equilibrium Supersonic Streams

1.1. *V-2 Automated Apparatus for Studying Molecular Spectra in Supersonic Gas Flows*

The experiments of this work were conducted on the V-2 apparatus, created in the Vibrational Division of the Institute of General Physics of the Academy of Sciences [21, 22] and consisting of three portions: a supersonic aerodynamic low-pressure tube, a submillimeter spectrometer, and a data acquisition and control system connected to a computer.

The gas-dynamic path (Fig. 1) serves to produce the supersonic flowstream of the gas mixture containing water vapor. It consists of a reservoir with water, which serves as the source of water vapor, a heater, nozzle, and vacuum chamber. The water vapor from the reservoir impinges on the mixing chamber through the shutter, then the gas (argon for example) passes through a device which regulates the flow rate, and then is mixed with water vapor. To improve the gas and water vapor mixing, they enter the mixing chamber through two openings situated in series. The jets flowing in series, owing to the expansion and turbulent nature of the flow, are mixed satisfactorily at a distance of a few tens of calibers. From the mixing chamber, the mixture impinges on the heater. The temperature of the mixture heater is controlled in the range of 290-430 K and is maintained constant within ±1 K using a computer. The heated mixture flows through the nozzle or slit, depending on the experiment, and flows into the vacuum chamber where a 1-10 Pa vacuum is maintained. The resulting supersonic flowstream is illuminated by a beam of submillimeter radiation in the 5-35 cm^{-1} frequency region. The vacuum in the chamber is maintained using pumps and a system of gas containers, and is measured by a manometer and a vacuum meter.

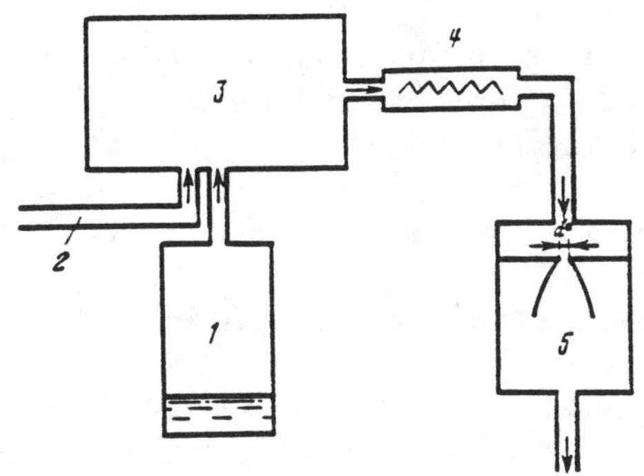

Fig. 1. Diagram of the V-2 instrument. *1*) water vapor source, *2*) inlet for the gas-carrier, *3*) mixer, *4*) heater, *5*) vacuum chamber.

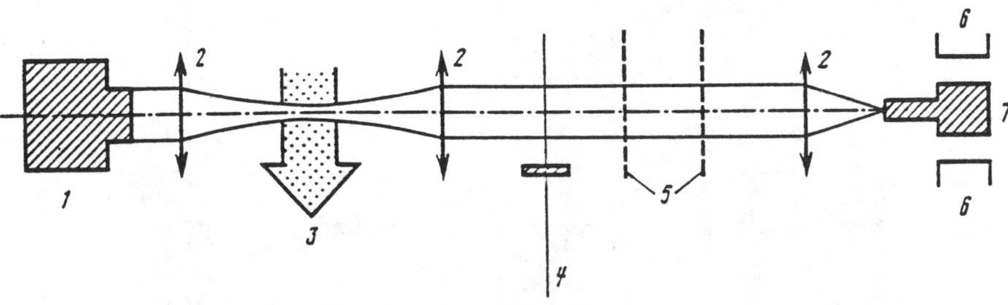

Fig. 2. Submillimeter spectrometer (diagram of the quasi-optical portion). *1*) OAP-5 acousto-optical transducer, *2*) Teflon lenses, *3*) supersonic flow, *4*) submillimeter radiation modulator, *5*) polarizers, *6*) electromagnet, *7*) monochromatic tunable submillimeter oscillator.

A diagram of the submillimeter spectrometer is shown in Fig. 2. A backward-wave tube type monochromatic generator with electronically tunable frequency is used as the submillimeter radiation source. The backward-wave tube is placed between the poles of an electromagnet with a magnetic field intensity of 5-13 kOe. The radiation emanating from the open end of the BWT-waveguide is focused using teflon lenses into a weakly diverging quasi-optical beam. After passing through the polarizer and modulator, the radiation enters the vacuum chamber through teflon lenses that are transparent to the BWT radiation. After transilluminating the supersonic flowstream, the radiation impinges on the active area of an OAP-5 acousto-optical transducer. The radiation frequency is adjusted by changing the anode voltage on the BWT. The range of anode voltage variation required to record a spectral line is 1-4 V and is determined by the gas pressure. The modulated signal from the transducer is fed to a synchronous detector.

The output signal from the synchronous detector is fed to the *X*-input of a dual-axis chart recorder and a voltage that is proportional to the anode voltage is fed to the *Y*-axis. In this fashion, as the anode voltage is changed, the recorder graphs of the variation of the intensity of the radiation passing through the chamber as a function of frequency.

The power supply for the BWT is made from a highly stable continuously tunable source developed in the Central Bureau for Unique Instrument Construction of the USSR Academy of Sciences.

An "Elektronika 60" minicomputer is used as the central core of the data acquisition and control system. A KAMAK programmable controller is used to connect the operating instruments with the transducers.

A block diagram of the data acquisition and control system is shown in Fig. 3. The signals fed to the inputs of the two-axis chart recorder are also fed to the digital voltmeters of the KAMAK system.

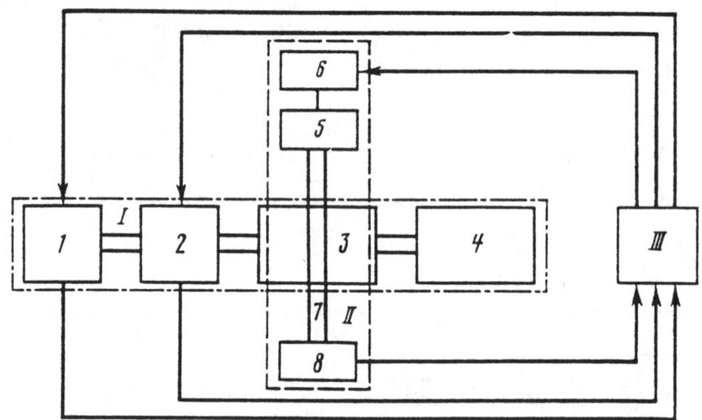

Fig. 3. Block diagram of the V-2 apparatus. *I*) low pressure aerodynamic tube: *1*) water vapor source, *2*) heater, *3*) vacuum chamber, *4*) vacuum reservoir; *II*) submillimeter spectrometer: *5*) BWT, *6*) stabilized power supply for the BWT, *7*) quasi-optical path, *8*) acousto-optical transducer; *III*) computer with data acquisition and control system.

The voltage measurement data are processed using the computer and the results are sent to the printer and the display. The processed results and the experimental points can be observed on the screen of an oscillograph. To this aim, a system is used that converts the numerical values into voltages connected to the oscillograph input. This same system serves to establish the required gas heating temperature. To close the relay (triggering the mechanism for changing the BWT's stabilized power supply anode voltage, and also the channels in the system that regulate the expansion of the gases) a relay control register (RCR) was used. The program for performing the experiment was written in "Extended BASIC", which is very suitable for a high-speed measurement program.

1.2. *Determining the Parameters of a Supersonic Flow of Rarefied Water Vapor by Quantitative Sub-Millimeter Spectroscopy*

Spectroscopic methods for determining the parameters of a moving gaseous medium are widely used in gas-dynamics. This section describes the adsorption method for measuring the temperature and density of a supersonic flowstream. The advantage of this method consists in the fact that a low-intensity transilluminating beam does not introduce any perturbation to the flow. The method's drawback is that the parameters are averaged over the length that the radiation interacts with the flow, it is therefore only suitable for measuring flowstreams with planar or cylindrical configurations.

To determine the temperature and density of the flowstream, the absorption corresponding to the transition between levels 2_{02} and 2_{11} of the rotational spectrum of water molecules ($\lambda = 0.398\ \mu m$) was measured. The range of the measured parameters was: water molecule concentration per unit volume $N = 10^{16} - 10^{17}\ cm^{-3}$, temperature 200 - 300 K, pressure 0.13 - 4 kPa at the measurement point.

The upper boundary in density and the lower boundary in temperature are determined by the onset of bulk condensation of water vapor in the flowstream.

1.3. *Theory of the Spectroscopy Diagnostic Method for a Supersonic Gas Flow*

Determination of the temperature T and static pressure P of a gas in a flowstream is based on measuring the absorption index α at the maximum of the spectral line contour, the width 2γ of the spectral line, and the total absorption index I of an isolated spectral line:

$$I = \int_{-\infty}^{\infty} \alpha(\omega - \omega_{21})\, d\omega, \tag{1}$$

where $\alpha(\omega - \omega_{21})$ is the absorption index as a function of frequency ω and ω_{21} is the frequency corresponding to the center of the spectral line.

The conditions indicated above correspond to homogeneous broadening caused by collisions. The absorption line shape is described by the function [23]

$$\alpha_T(\omega) = \frac{4\pi^2\mu^2 R}{3\hbar c}\omega G(\omega - \omega_{21})(n_1 - n_2), \tag{2}$$

$$G(\omega - \omega_{21}) = 1/\pi\nu_c[1 + (\omega - \omega_{21})^2/\nu_c^2],$$
$$n_i = N_i/g_i, \quad i = 1, 2, \tag{3}$$

where ν_c is the mean collision frequency, μ = 1.8546 ± 0.0006 D is the dipole moment of $H_2{}^{16}O$ [3], R = 2.0757 is the spectral line strength for the 2_{02}-2_{11} transition [24], c is the speed of light, $\hbar$ is Plank's constant, N_i is the population, and g_i is the degeneracy of the ith level.

From the form of the theoretical function for $\alpha_T(\omega)$, it follows that the experimentally observed absorption index is described by a function of the form

$$\alpha_э(\omega) = \frac{\omega}{\omega_{21}} \frac{\alpha}{1 + (\omega - \omega_{21})^2/\gamma^2} \approx \frac{\alpha}{1 + (\omega - \omega_{21})^2/\gamma^2}. \tag{4}$$

Here 2γ is the full-width of the spectral line at half-maximum. For this function, the total absorption can be calculated easily from Eq. (1):

$$I = \pi\alpha\gamma. \tag{5}$$

On the other hand, the theoretical value of the total absorption index can be obtained by integrating Eq. (2):

$$I_T = \sigma\omega_{21}(n_1 - n_2), \tag{6}$$

where σ is a quantity having the dimensions of a cross-section and composed of constants characterizing the molecule and the spectral transition. For our case, $\sigma = 2.973 \cdot 10^{-18}$ cm^2.

Setting Eqs. (5) and (6) equal to each other, we obtain an expression that relates the parameters of the spectral line α, γ with the value of n_1 - n_2, which is a function of the concentration of water molecules and the gas temperature:

$$n_1 - n_2 = \pi\alpha\gamma/\sigma\omega_{21}. \tag{7}$$

In this section, we shall examine only those gas flows which satisfy the condition of rotational equilibrium in the measurement region. This means that the rotational level distribution of water molecules is a Boltzmann distribution in the place where the gas flow parameters are determined. In this case

$$n_1 - n_2 = N \frac{g_I}{Q(T)} (e^{-\epsilon_1/T} - e^{-\epsilon_2/T}), \tag{8}$$

where ϵ_1 and ϵ_2 are energy levels, $Q(T)$ is the rotational statistical sum, and g_I is the nuclear spin degeneracy. For our case, ϵ_1 = 100.93 K, ϵ_2 = 137.05 K, and g_I = 1. For $H_2{}^{16}O$ water molecules at a temperature above 140 K, it is valid to calculate the value of $Q(T)$ using the formula [23-25]

$$Q(T) = 2\sqrt{\pi T^3/ABC}, \tag{9}$$

where A = 27.876 cm^{-1}, B = 15.507 cm^{-1}, C = 9.288 cm^{-1} are the rotational constants for the vibrational ground state [26]. The product of the three rotational constants can be represented as the cube of some arbitrary temperature T = 22.384 K.

In the majority of gas-dynamics problems, not only do the values of N and T vary, but so does their relationship to the stagnation parameters: N/N_0 and T/T_0. For this to be applicable to the flow in supersonic flowstreams, it is necessary to once again measure the parameters of the spectral line α_0 and γ_0 in the motionless gas. Then α and γ are measured under supersonic flow conditions for a gas with stagnation parameters N_0 and T_0. Using Eqs. (7)-(9), we obtain

$$\frac{\alpha\gamma}{\alpha_0\gamma_0} = \frac{NQ(T_0)}{N_0Q(T)} \frac{e^{-\epsilon_1/T} - e^{-\epsilon_2/T}}{e^{-\epsilon_1/T_0} - e^{-\epsilon_2/T_0}} \approx \frac{N}{N_0}\left(\frac{T_0}{T}\right)^{5/2} \exp\left(\frac{\epsilon_1}{T_0} - \frac{\epsilon_1}{T}\right), \tag{10}$$

where $(\epsilon_1 - \epsilon_2)/T \ll 1$ has been taken into account.

The right hand side of Eq. (10) is a function of two variables of interest to us: N and T. It is possible to eliminate one of these by several means. The first approach consists in an assumption regarding the nonisentropic nature of supersonic flow from the onset of acceleration to the measurement point. In this case, N and T are related by the adiabatic equation

$$N/N_0 = (T/T_0)^{1/(\kappa-1)}, \tag{11}$$

where κ = 4/3 is the adiabatic constant for water vapor. Substituting Eq. (11) into Eq. (10), we get

$$\alpha\gamma/\alpha_0\gamma_0 = (T/T_0)^{1/(\kappa-1)-5/2}\exp(\epsilon_1/T_0 - \epsilon_1/T). \tag{12}$$

By experimentally determining the left-hand-side of Eq. (12), we can find T by numerically solving Eq. (12).

The second means consists in using the dependence of the line-width 2γ on temperature and density $\gamma = \gamma(N, T)$. This function for the lines of the rotational spectrum of water molecules, lying in the millimeter and submillimeter regions, has been measured repeatedly [27]. It is represented by the formula

$$\gamma/\gamma_0 = (N/N_0)\,(T/T_0)^{1-n}, \tag{13}$$

where the exponent $n \sim 0.5$-1.5. Substituting Eq. (13) into Eq. (10), we also obtain an equality which contains only the temperature as an unknown:

$$\alpha/\alpha_0 = (T/T_0)^{n-7/2}\exp(\epsilon_1/T_0 - \epsilon_1/T). \tag{14}$$

The third means consists in measuring the total absorption index for two spectral lines. In this case, from Eq. (10) we get

$$\frac{\alpha\gamma}{\alpha_0\gamma_0}\,\frac{\alpha_0'\gamma_0'}{\alpha'\gamma'} = \exp\left[(\epsilon_1 - \epsilon_1')\left(\frac{1}{T_0} - \frac{1}{T}\right)\right]. \tag{15}$$

It is obvious that for precise measurements, the values of ϵ_1 and ϵ_2' should not be close. From a practical point of view, sufficient accuracy is provided by the condition $\epsilon_1' - \epsilon_1 \sim T$. The latter method is the most universal, since it is related to a single assumption regarding the rotational equilibrium of the gas flow.

From here it follows that if the equilibrium temperature can be determined by some independent means, then, by calculating T from Eq. (15), it is possible to judge the degree of nonequilibrium in relation to the rotational motion. As a consequence, Eq. (15) will be used iteratively to analyze the character of the molecular rotational level distribution. Here, we shall also examine only the first two means illustrating the determination of the temperature inside a planar Laval nozzle.

1.4. *Order of Measurements and Processing the Experimental Data*

The standard measurement algorithm was chosen to be as follows:

1. A voltage across the BWT is established such that the wavelength of the radiation is close to the studied absorption line.
2. A slowly varying voltage on the BWT anode is continuously measured; this voltage causes a change in the frequency of the radiation [28]. Also, at 120 preassigned points uniformly distributed with respect to the voltage, readings of the microwave radiation power are taken at the exit of the OAP-5 transducer.
3. When a given value of the BWT voltage is reached, the measurements are stopped and the system is returned to its initial state. The initial voltage is chosen such that the absorption line is located approximately in the middle of the covered frequency range.

Two approaches are taken in each experiment: one, when blowing gas that does not contain any water vapor through the nozzle to obtain a precise amplitude-frequency response of the spectrometer and, the second, when running a flow of water vapor through the nozzle. Two blocks of 120 numbers are obtained ($J = 1, ..., 120$), corresponding to the readings of the radiant power passing through the studied flow with and without water vapor: $U_1(J)$ and $U_2(J)$.

Under the experimental conditions for observing the 2_{02}-2_{11} line, the tuning range of the anode voltage was a small fraction (1 V) of the voltage on the BWT anode (3 kV), such that the frequency of the BWT radiation could be treated as a linear function of the anode voltage and the frequency measurements could be replaced by electrical measurements.

The task of determining the parameters of the line α, γ consists in plotting by the method of least squares a curve of the form of Eq. (4) through the experimental points α_e. The absorption index $\alpha_e(J)$ for the Jth measurement points is first calculated from the equation

$$\alpha_e(J) = \ln[U_2(J)/U_1(J)].$$

We shall rewrite Eq. (4) in the form

$$1/\alpha_e = 1/\alpha + (J - J_0)^2/\alpha L^2; \qquad (16)$$

in this equation, the frequency intervals are replaced by the number of measurement points present in this interval, L is the number of points present in the interval γ, and J_0 is the point corresponding to the maximum absorption index.

Fig. 4. Flow chart for measuring and processing the experimental data. $S = 0$ pertains to measurement with water vapor, $S = 1$ is without it.

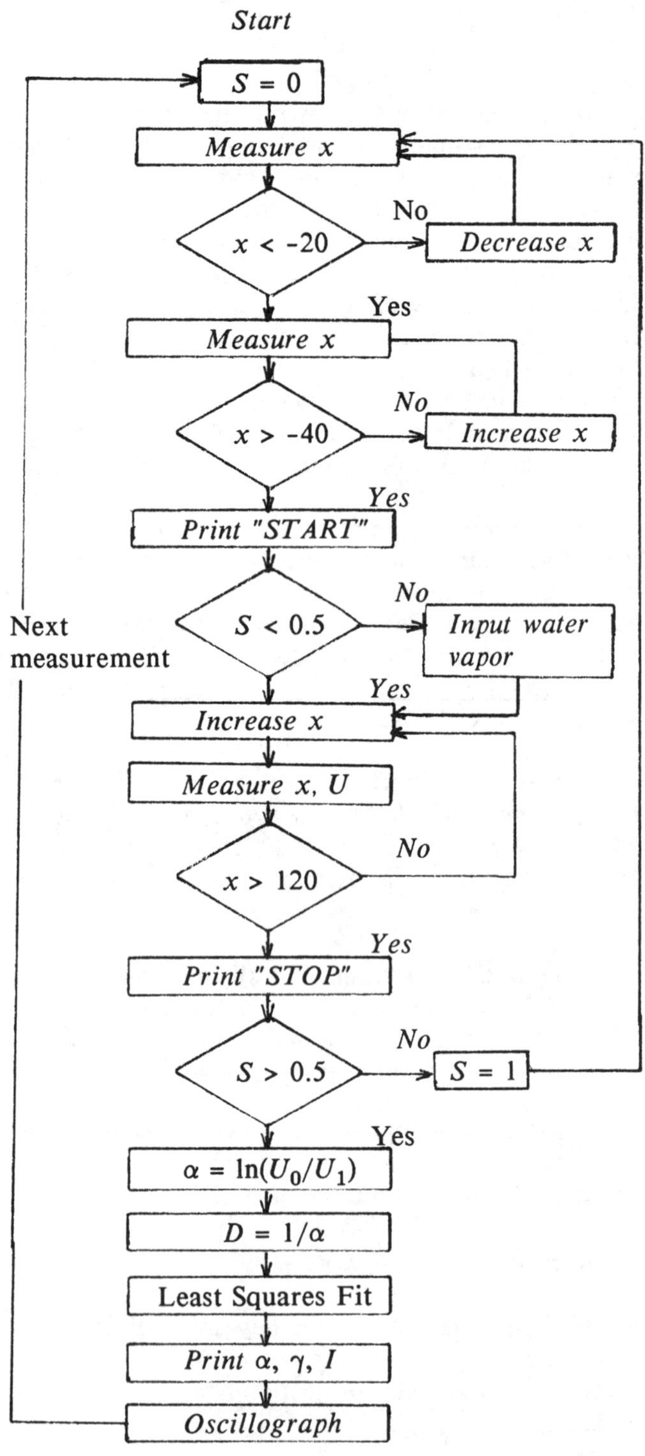

Equation (16) is advantageous in that the block of experimental points for $1/\alpha_e(J)$ is described by a polynomial that is second order in J. Finding the coefficients of the polynomial by the least-squares method (LSM) is a well-known procedure that is described in many text books (see, for example, Ref. [29]). It reduces to solving normal linear equations determining the coefficients of the polynomial. The resulting values for the coefficients enable one to calculate the parameters of the spectral line.

A flow chart for the measurement program is shown in Fig. 4.

1.5. *Results of Measurements in Planar Wedged Supersonic Streams*

Measuring the parameters α, γ of the 2_{02}-2_{11} spectral line in a supersonic flowstream of water vapor was done for two planar tapered nozzles, data on which are given in Table 1. Also given are the calculated temperatures for $\kappa = 4/3$ with and without allowance for the boundary layer.

According to theory, for the shock mechanism of spectral line broadening, the frequency of the broadening collisions ν_c and the line-width γ are directly proportional to the density of the gas, and the absorption index at maximum $\alpha_T(\omega_{21})$ does not depend on the gas density. Experimental confirmation of these relationships is a good test of the absence of systematic errors in spectral measurements.

Figure 5 shows the absorption index α at maximum and the total absorption index $I = \pi\gamma\alpha$ as functions of pressure P_0 for $T_0 = 288 \pm 0.5$ K for measurements in flowing and motionless gases. The points in Fig. 5, *a* within $1000 < P_0 < 2600$ Pa are situated parallel to the abscissa axis, and in Fig. 5, *b* are located within experimental error (±3%) on lines that pass through the origin, which is consistent with theoretical predictions. The deviation from the theoretical dependence for $P_0 > 2600$ Pa for nozzle 2 on both graphs is explained by bulk condensation of water vapor.

Table 1

Technical characteristics of the nozzles and temperatures determined by different methods

Characteristic	Nozzle		Temperature	Nozzle	
	1	2		1	2
h	8.3	8.4	$T(12)$	241 ± 1	225 ± 1
l	90.0	90.0	$T(14)$	241 ± 1	222 ± 1
x	50.0	50.0	T_1	226	209
H	8.8	10.0	T_2	226	
M	1.28	1.51			

Note: h is the height of the critical cross-section of the nozzle in mm, l is the width of the critical cross-section or the extent of the absorption zone in mm, x is the distance from the critical cross-section to the observation point or the center of the microwave beam in mm, H is the height of the nozzle at the observation point in mm, M is the calculated Mach number at the observation point; $T(12)$ and $T(14)$ are the temperatures determined from Eqs. (12) and (14), respectively, in K; T_1 and T_2 are the calculated temperatures with and without allowance for the boundary layer, respectively, in K.

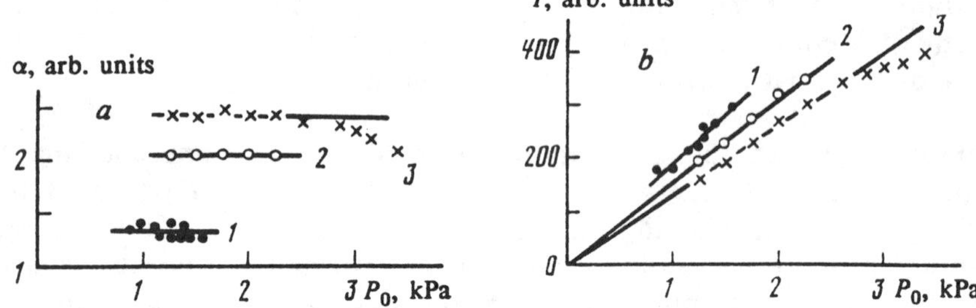

Fig. 5. Variation of the absorption index α at the center of a Lorentz contour and the total absorption I as a function of initial water vapor pressure P_0 for flows in nozzles and in a motionless gas. *1*) motionless gas; *2*) nozzle 1; *3*) nozzle 2.

For the experiments, we selected tapered nozzles with a small divergence angle in the supersonic portion, the gas flow in which was isentropic to a high level of accuracy. In this case it was possible to compare both methods of determining the supersonic flow temperature.

In the first method, the flow is assumed to be isentropic. Then from the ratio of the total absorptions in Eq. (12), the value of the temperature T can be found. In the second method, the flow is not assumed to be isentropic, but the variation of the line-width with temperature is assumed to be known (Eq. (13)) and the value of T is found from the ratios of the absorption indices at maximum. The index n for the 2_{02}-2_{11} line was specially measured in the 220-290 K temperature range and turned out to be equal to 0.90 ± 0.05.

Table 1 cites the values of T determined by the two methods. These values coincide within experimental errors, which attests to the internal consistency of the methods, since the assumptions in the first case were related to the mechanics of a continuous medium, and in the second case, were related to the area of molecular spectroscopy. It is interesting to compare the values of the temperature obtained by the spectroscopic method with that which modeling the flow in jets in the one-dimensional variant yields.

As is clear from Table 1, the temperature for the observation point, calculated without allowing for the boundary layer, is 6% less than that obtained in the experiment. In order to account for the thickness of the boundary layer, we shall make use of an equation from Ref. [30]

$$\frac{\delta^*}{x} = \frac{1.74}{\sqrt{\mathrm{Re}_x}}\left(1 + \frac{48}{35}\,\frac{\kappa - 1}{2}\,M^2\right)\left(1 + \frac{\kappa - 1}{2}M^2\right)^{\frac{a-1}{2}},$$

where δ^* is the thickness of the displacement layer, Re_x is the Reynolds number, M is the Mach number in the flow, and a is the temperature index of the viscosity coefficient. The values of the viscosity coefficient were taken from the tables in Ref. [31]. Allowing for the boundary layer, the calculated temperature coincided within experimental error with that determined by the spectroscopic method (see Table 1).

The spectroscopic method, consisting in measuring the line-width, also permits us to determine another variable of interest - N - whose value is calculated from Eq. (13) and for an isentropic flow is calculated from Eq. (11). The proposed method of determining the temperature of supersonic flows permits generalization to that case where the water vapor is a small impurity in the composition of the gas mixture. An application can be found for this when studying complex supersonic flowstreams, when the calculation methods for determining the flow parameters become unreliable.

2. Rotational Nonequilibrium of $H_2{}^{16}O$ Molecules in a Supersonic Jet of Rarefied Water Vapor

The measurements described in Section 1 showed that the thermodynamic parameters of a gaseous medium, determined under equilibrium supersonic flow conditions using Eqs. (1)-(7) relating the spectral parameters α and γ to the population difference $n_1 - n_2$, agree within experimental error with the calculated parameters determined using gas-dynamic calculations. In essence, this means that the population difference $n_1 - n_2$, calculated from Eq. (8) for equilibrium values of gas flow parameters N and T, coincide at equilibrium with the value of $n_1 - n_2$ determined from spectroscopic measurements by Eq. (7). It is obvious that for nonequilibrium conditions, generally speaking, there will be no such correspondence, and the ratio $K = (n_1 - n_2)/(n_1^e - n_2^e)$ will characterize the nonequilibrium of the rotational distribution. From Eq. (7) it follows that $n_1 - n_2$ is proportional to $\pi\alpha\gamma = I$. By determining the equilibrium total absorption I^e using Eqs. (7) and (8), we get

$$K = (n_1 - n_2)/(n_1^e - n_2^e) = I/I^e. \qquad (17)$$

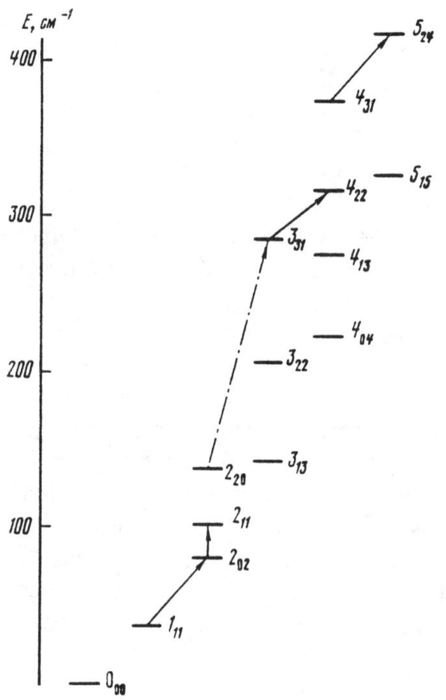

Fig. 6. Diagram of the lower energy levels of the $H_2{}^{16}O$ molecule. The arrows denote transitions for which the absorption coefficient was measured.

If it is possible to calculate the equilibrium values of N and T for the experimental conditions, then, by measuring the total absorption, it is possible to determine the level of nonequilibrium using Eq. (17) (for equilibrium $K = 1$).

Experiments shall be described in this section which were conduced in a planar supersonic jet of rarefied water vapor. It is well known that nonequilibrium homogeneous condensation is observed under certain initial conditions in supersonic jets [32].

If we move downstream from the critical section, it is possible to observe sequentially a region of flow without condensation and a region where nonequilibrium condensation begins, and a zone of well-developed condensation.

These zones can also be observed if the distance from the critical cross-section does not change. For this to occur, it is necessary to change the initial temperature within a certain range of values at constant initial density of water vapor.

Table 2

Spectral characteristics of rotational transitions

Transition	E_1, cm^{-1}	E_1-E_2, cm^{-1}	R	$\sigma \cdot 10^{18}$, cm^2
1_{11}-2_{02}	37.137	32.951	0.7549	1.0807
2_{02}-2_{11}	70.088	25.085	2.0752	2.9710
3_{31}-4_{22}	285.217	30.560	0.1615	0.2312
4_{31}-5_{24}	383.837	32.365	0.2619	0.3750

Note: E_1 and E_2 are the lower and upper levels of the transition, R is the intensity of the rotational transition line (see Eq. (2)), and $\sigma = 4\pi\mu^2 R/3\hbar c$ (see Eq. (6)).

According to the calculations of Ref. [14], in the region corresponding to the onset of nonequilibrium condensation, there is a large number of unstable, i.e. vibrationally-excited, water molecule associates $(H_2O)_n^*$. The excited complexes transfer their energy to the rotational degrees of freedom of the H_2O molecules, which should distort the equilibrium rotational energy level distribution. The transfer of energy is most effective in the resonant case, when the frequency of the vibrations coincides with rotational transition frequency. In Ref. [23] when studying the rotational spectra of water dimers $(H_2O)_2$, the vibrational frequency of the dimer was determined at which the distance between H_2O molecules changes. It is roughly 150 cm^{-1} and almost coincides with the frequency of the 2_{20}-3_{13} transition in the H_2O rotational spectrum (Fig. 6). When converting vibrational energy into rotational, one should expect an increase in the population at the 3_{13} level and a decrease at the 2_{20} level relative to equilibrium. Therefore, for the 3_{13}-4_{22} transition, in which the 3_{13} level is the lower, an increase in the coefficient K can be expected for low initial water vapor temperatures. To test these propositions, the total absorption was measured in a supersonic jet of water vapor for the transitions 1_{11}-2_{02}, 2_{02}-2_{11}, 3_{31}-4_{22}, 4_{31}-5_{24} for different T_0 (see Fig. 6 and Table 2).

2.1. *Method for Modeling the Flow of a Supersonic Stream of a Gas and the Experimental Conditions*

The distribution of the equilibrium gas parameters (molecular concentration N and gas temperature T) necessary to determine the value of K were calculated using well-known parameters of the initial state of the gas (N_0, T_0). A stationary two-dimensional calculation technique was used to model the supersonic flow behind the acoustic nozzle. The technique is based on using a "natural" coordinate system formed by the flow lines and the normals to the flow lines [34, 35]. In this method, the entire flow field is decomposed into a series of flux tubes, the flow in each tube is calculated in a one-dimensional representation, and the tubes interact normal to their boundary through pressure and via the velocity derivative of the pressure with respect to the turning angle of the tube. This calculation technique permits one to allow for the rotational relaxation of molecules as well has homogeneous gas condensation in the jet [35].

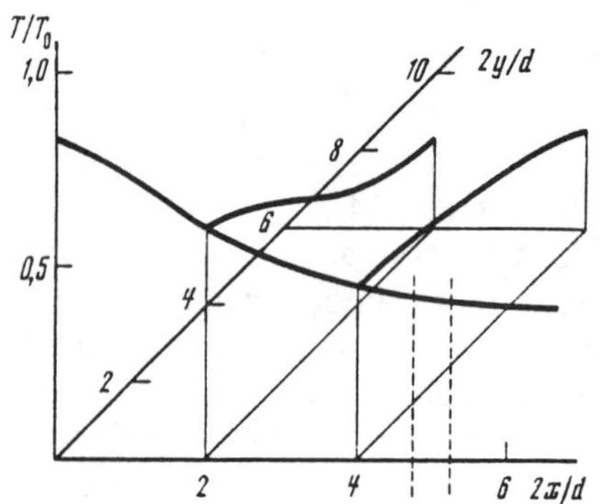

Fig. 7. Temperature distribution inside a planar two-dimensional jet. The OX-axis coincides with the direction of the gas motion, the OY-axis is perpendicular to the symmetry plane of the jet, the origin of the coordinate system is located on the slit of the supersonic nozzle, and d is the height of the critical nozzle.

Modeling the gas flow in a planar freely-expanding jet was done assuming isentropic flow with $\kappa = c_p/c_V = 1.33$. This value of κ corresponds to assuming gas flow in rotational equilibrium for a gas consisting of molecules with three rotational degrees of freedom. Figure 7 shows the variation of T/T_0 with the x-coordinate along the central flow line, and also the profile of T/T_0 along the transverse coordinate y for several distances referenced from the slit of the acoustic nozzle. The position of the center of the microwave beam is characterized by the parameter $2x/d$ = 5.5, where the height of the critical cross-section is d = 12 mm. The boundary of the beam at the half-intensity level is shown by the dashed line. For the center of the beam, T/T_0 = 0.4015 and N/N_0 = 0.06472. The nonuniformity with respect to T/T_0 within the beam over the two coordinates did not exceed 8%.

The initial water vapor pressure P_0 = 533 Pa was identical in all experiments. Since the gas flowed into the vacuum chamber at a constant pressure of 13 Pa, the supersonic gas flow took place in a confined region of space.

The range of initial temperatures T_0 = 293 - 423 K was chosen such that the it would be possible to shift the zone of nonequilibrium condensation along the jet relative to the observation point.

According to the calculations in Ref. [35], for 293 < T_0 < 323 K, the observation point is located in a region of well-developed condensation, for 323 T_0 < 373 K it is in the zone where condensation just begins, and for T_0 > 373 K it lies within a flow region without condensation.

2.2. *Results of Measurements in Planar Supersonic Jets of Water Vapor at Different Stagnation Temperatures*

The rotational spectrum of an H_2O molecule is decomposed into two systems of levels, one of which corresponds to total nuclear spin 0 (*para*-modification) and the other to spin 1 (*ortho*-modification). The radiative transitions between levels of different spins are forbidden by the selection rules, whereas collisional transitions are unlikely under normal conditions due to the exclusion on nuclear spin. The rotational levels of all transitions for which measurements were made (see Table 2) are referenced to the *para*-modification. The energy intervals dividing the pairs of rotational levels for these transitions were almost identical (25-33 cm^{-1}).

Figure 8 shows the results of the measurements in the form of the dependence of the coefficient K on the initial water vapor temperature T_0, which shows that a supersonic jet of rarefied water vapor is nonequilibrium to a strong degree in relation to the rotational degrees of freedom of H_2O molecules. The nonequilibrium is especially strong for low initial temperatures under condensation conditions.

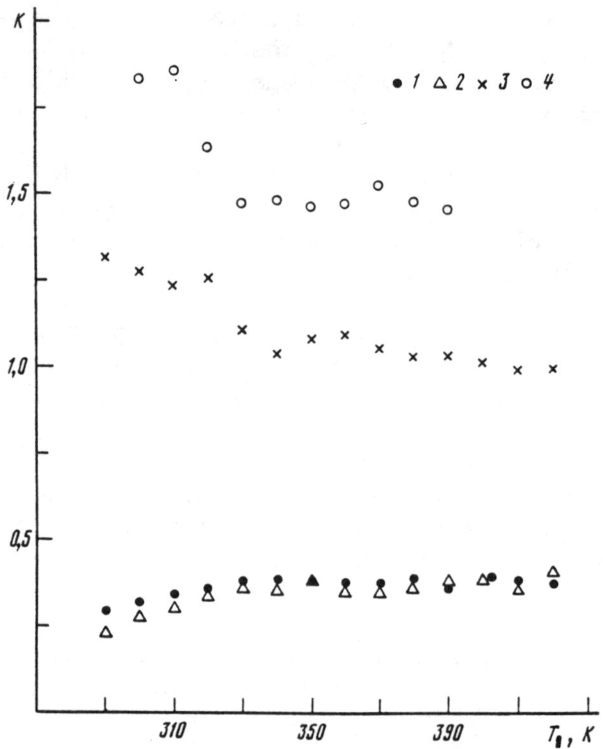

Fig. 8. Nonequilibrium coefficient K as a function of initial water vapor pressure. *1*) 1_{11}-2_{02} transition, *2*) 2_{02}-2_{11}, *3*) 3_{31}-4_{22}, *4*) 4_{31}-5_{24}. $K = 1$ in the case of rotationally equilibrium flow.

For the 3_{31}-4_{22} transition, the function $K(T_0)$ is in agreement with the proposed variation: for high temperatures $K \approx 1$, which corresponds to an equilibrium distribution in the absence of condensation. When the initial temperature T_0 is lowered, the coefficient K increases beginning at the value of T_0 for which condensation should occur at the observation point according to calculations. For other transitions, the levels of which are situated above levels 3_{31} and 4_{22} (4_{31}-5_{24}) or below them (1_{11}-2_{02} and 2_{02}-2_{11}), the absorption is significantly different from the equilibrium value even for such T_0 at which condensation is not possible.

This can be explained by the emerging rotational nonequilibrium being related not only to condensation, but also to the small rotational relaxation rate. The energy interval ΔE for rotational levels between which the molecular collisional transitions take place reaches 80-150 cm^{-1} for several transitions in the spectral region being studied. On the other hand, the average temperature in the jet is 150 K (or 100 cm^{-1}), i.e., $\Delta E_{ro} \sim T$, therefore the rotational relaxation rate may be insufficient to establish equilibrium under the conditions in a supersonic jet.

The nonequilibrium distribution, related to the finite rotational relaxation time, was observed in a supersonic jet of molecular nitrogen for high levels J = 8-12 [11], and also under conditions of a gas-dynamic molecular beam source for linear OCS molecules, where the deviation from equilibrium was recorded at low rotational levels J = 1-3 [13]. For a gas consisting of polyatomic polar asymmetric rotator molecules, rotational nonequilibrium in supersonic flowstreams had not been observed before.

When describing a rotationally nonequilibrium state, the concept of rotational temperature is often used, when it is assumed that the rotational level distribution is Boltzmann, but the distribution parameter T_{ro} differs from the gas temperature. We shall show that in our case the molecules' rotational level distribution can not be described using a Boltzmann distribution function.

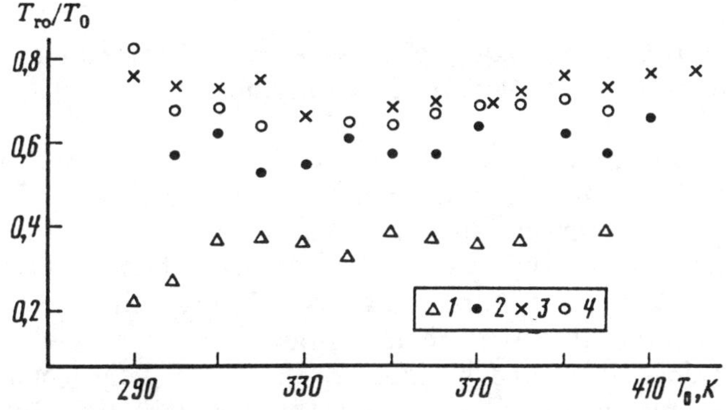

Fig. 9. Rotational temperature as a function of initial water vapor temperature determined for different pairs of rotational transitions.
1) 1_{11}-2_{02} and 2_{02}-2_{11}, *2*) 3_{31}-4_{22} and 4_{31}-5_{24}, *3*) 1_{11}-2_{02} and 3_{31}-4_{22}, *4*) 1_{11}-2_{02} and 4_{31}-5_{24}.

In Section 1, it was shown that the rotational temperature distribution can be determined from the ratio of the total absorption at two different transitions using Eq. (15). Figure 9 shows values of T_{ro}/T_0 calculated for different pairs of rotational levels as functions of the initial water vapor temperature T_0. If the concept of rotational temperature were valid for the observed distribution, then the variation of T_{ro}/T_0, determined for different combinations of rotational level pairs, would fall within experimental errors. For example, if the flow is completely equilibrium, then all points are grouped around the horizontal line $T_{ro}/T_0 = 0.4$; if the flow is completely stagnant, then $T_{ro}/T_0 = 0.9$. Since the value of T_{ro}/T_0 is spread in the interval of 0.2-0.8, then it turns out to be impossible to introduce the rotational temperature. Note that in the absence of condensation, the value T_{ro}/T_0 determined for the 1_{11}-2_{02} transition is close to 0.4, which corresponds to the equilibrium value.

So, as a result of the experiments, it turned out that:

1) the rotational level distribution is strongly nonequilibrium,

2) the nonequilibrium is preserved when the temperature is increased to values at which condensation effects can be neglected ($T_0 > 373$ K),

3) when the temperature is increased above 333 K, the values of the nonequilibrium coefficients and T_{ro}, calculated for different pairs of rotational levels, are almost invariant,

4) the distribution can not be considered to be Boltzmann either at such T_0,

5) two transitions exist for which T_{ro}/T_0 is close to equilibrium and the population difference for each is 0.4 of the equilibrium population.

In Ref. [15], two suggestions were made regarding the reasons causing the rotationally nonequilibrium distribution.

The first suggestion was based on the fact that the water vapor parameters are close to the saturation line and the flow is condensation-nonequilibrium. This means that the concentration of small formations of the dimer and trimer type, and also large size clusters is less than the equilibrium value. The formation and decomposition of clusters is accompanied by the liberation or absorption of energy much greater than the average translational and vibrational energy of a free molecule. In a condensation-nonequilibrium gas, the rates of the particle formation and break-up processes are different. If we assume that the exchange of energy during formation and break-up takes place not only through the translational degree of freedom, but also involves the rotational degree of freedom, by allowing for the finite rate of rotational relaxation, one can expect the appearance of the rotational nonequilibrium effect.

This mechanism, apparently, plays in important role at low (< 333 K) water vapor temperatures T_0. However, using this mechanism, it is difficult to explain the fact that the nonequilibrium coefficient K for $T_0 < 333$ K is almost independent of temperature for all transitions. Meanwhile, the concentration of water vapor associates should drop with increasing temperature. Therefore, a second proposal was put forth consisting in the nonequilibrium rotational distribution being a result of the low rate of rotational relaxation, and the observed distribution is some stage of a transient process from the initial temperature T_0 to the equilibrium gas temperature T of the jet.

This suggestion makes it possible to explain fundamental details of the nonequilibrium rotational distribution for high T_0. For example, point 5 of the conclusions can be treated as a result of the relaxation rate depending strongly on the size of the energy interval ΔE that separates a pair of rotational levels.

The rotational spectrum can be arbitrarily divided into adiabatic and nonadiabatic regions. This separation for water molecules at 150-300 K temperatures corresponds to 30-50 cm^{-1} and is determined by the relationship between the angular velocity relative to the motion of the molecules and the frequency of the rotational transition (see, for example, Ref. [36]). For the majority of transitions of the H_2O molecular rotational spectrum, the frequencies lie above this limit. The only exception is the group of the lowest levels.

During relaxation from high initial temperatures T_0 to low final temperatures, the population of the lower levels, for which $E_{ro} < T$, should increase due to the transition from the upper levels ($E_{ro} > T$). But for the region of the rotational spectrum $E_{ro} > T$, the energy intervals between the levels are great and, consequently, the relaxation rates are small. Therefore, the particles are confined to this region of rotational energies, which generates a deficiency of the total number of particles in the group of lower levels. Meanwhile, for the group of lower levels the relaxation rate is sufficient to establish equilibrium in relation to the gas temperature T of the water vapor jet.

For the 4_{31}-5_{25} transition, associated with the upper region of the spectrum of rotational energies, the population of the levels exceeds the equilibrium population and the coefficient $K > 1$.

The 3_{31}-4_{22} transition belongs to the intermediate energy region of the rotational energy spectrum, where the populations change little with a change in temperature from T_0 to the equilibrium temperature T of the gas in the observation region; therefore, for this transition $K \approx 1$.

To obtain quantitative values of the quantities describing the relaxation, and for more definite conclusions regarding the picture of relaxation as a whole, additional research is necessary and, to the extent possible, for a larger number of transitions. This is difficult to accomplish for water molecules, since for the majority of its transitions ΔE lies at the limits of the range attainable by a BWT-tube. This difficulty can be overcome if the heavy isotope D_2O is used instead of water. It has a similar spectrum, but with almost twice as many spectral lines. An advantage to working with D_2O is that there is no absorption in atmospheric air for the lines of D_2O.

3. Nonequilibrium Rotational Distribution Function for D_2O Molecules in a Rarefied Supersonic Jet of Ar

3.1. *Choice of Initial Conditions for the Supersonic Jet*

The results obtained at the end of Section 2 show that a nonequilibrium distribution in the H_2O rotational spectrum is observed under supersonic jet conditions both with and without condensation. In order to clarify what contribution the rotational relaxation process itself brings to the overall nonequilibrium, it is necessary to eliminate water vapor condensation. This can be done either by raising the temperature T_0 for P_0 = const, as in Section 2, or by lowering the pressure P_0 for T_0 = const. If room temperature is chosen for T_0, then the second approach is technically simpler to implement.

However, by lowering P_0, we go from the region of continuous flow of the medium to an intermediate region, where the theoretical calculations for the flow of the jet, as was discussed above, are no longer applicable. In order to avoid the difficulties associated with the intermediate flow region and to fulfill the conditions for the flow of a continuous medium, a jet of argon was formed at a sufficiently high initial pressure P_0. To the argon we added D_2O vapor in such an amount that we could neglect condensation effects. In the resultant jet, a nonequilibrium rotational distribution of D_2O molecules was observed that was caused by insufficiently rapid relaxation of heavy water molecules. The observations were made in the plane of the supersonic argon jet. The concentration of D_2O was varied from very low ($< 1\%$) to 7.5%.

For small concentrations, it can be assumed that D_2O molecules collide only with Ar atoms and that D_2O-D_2O collisions can be neglected. When the concentration is increased to 5-7%, the number of D_2O molecules is still too small to have a noticeable effect on the gas-dynamic properties of the jet. However, D_2O-D_2O collisions have a much more marked effect on relaxation than do D_2O-Ar collisions, therefore the nonequilibrium distribution changes significantly. This permits one to compare the effect of both types of collisions on the rotational relaxation of D_2O molecules for almost identical rates of temperature variation and gas flow velocities.

A diagram of the variation of the absorption coefficient in a planar supersonic jet of Ar-D_2O is shown in Fig. 2.

The initial pressure and temperature of the gas were 266 Pa and 293 K, respectively. The calculated gas temperature in the range of variation was 65 ± 5 K. In order for the jet to be considered planar and to avoid spreading in the coordinate parallel to the slit, teflon walls were installed. For gas flow on the teflon walls, a boundary layer emerged, the absorption in which could be attributed to measurement errors. However, the temperature in the boundary layer was close to the stagnation temperature and, consequently, was 5 times greater than the temperature of the main flow and, hence, the density there was 5 times less. The measurements were done for low vibrational transitions, the absorption at which drops strongly with increasing temperature. For these reasons, the absorption in the boundary layer did not exceed 1-2% of the absorption in the center of the supersonic jet.

3.2. *Classification of the Rotational Levels of the* D_2O *Molecule*

We shall examine the diagram of the lower rotational levels of D_2O (Fig. 10). D_2O molecules can have integer nuclear spin equal to 0, 1, or 2. Molecules, having even nuclear spin belong to the *ortho*-modification and those with odd spin belong to the *para*-modification. Collisional and radiative transitions between different modifications are forbidden with a high degree of accuracy. Therefore, for the majority of experimental situations, the *para*- and *ortho*-systems can be treated as being independent. Figure 10 shows the rotational levels of the *ortho*-system of D_2O for which the measurements were conducted. The wave functions of the *ortho*-modification of D_2O are transformed along irreducible representations A and B_1 of group D_2, if the Oz'-axis of the movable coordinate system is directed along the symmetry axis of the D_2O molecule. With the quantization of the asymmetric rotator, the division into (+)- and (-)-functions arises. The (+)-functions do not change on reflection in a plane passing through the Oz'-axis of the moving coordinate system, whereas the (-)-functions change sign [37].

Recall that, in molecular spectroscopy, there is a division of the levels into positive or negative, depending on whether the wave function changes its sign when the coordinate system is inverted. In order to avoid confusion, we shall introduce the notations σ^+ and σ^- for the (+)- and (-)-functions, respectively.

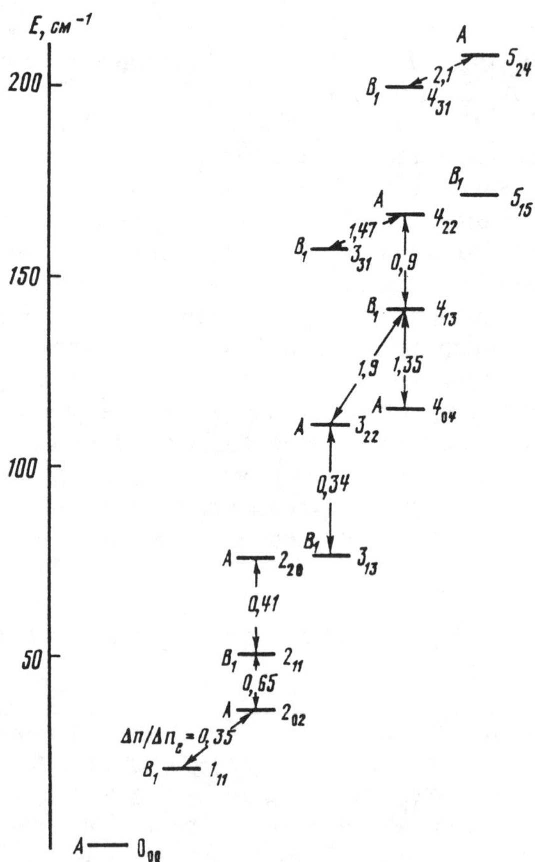

Fig. 10. Diagram of the lower rotational energy levels of molecular D_2O (*ortho*-modification).

3.3. *Experimental Results and their Qualitative Interpretation*

In the experiments, the total absorption was measured for the transitions indicated in by the arrows in Fig. 10. The total absorption, as follows from Eq. (7), is proportional to the population difference Δn. For the conditions under which our observations were made, it is possible to calculate the equilibrium total absorption, which is proportional to the equilibrium population difference Δn_e. The ratio of the absorption to the equilibrium absorption $\Delta n/\Delta n_e$ is shown in Fig. 10 next to the arrows. Attempts to describe the experimental results of measuring the total absorption using a Boltzmann distribution at some temperature were unsuccessful. This is usually quite evident when comparing $\Delta n/\Delta n_e$ for the transitions 3_{13}-3_{22} and 3_{22}-4_{13}, having level 3_{22} in common. For the first transition we have a population difference of 0.34 from equilibrium, and for the second we have 1.9, which is impossible for any single positive temperature for the entire system of energy levels.

This can be explained by the accumulation of particles at level 3_{22}, which is related to the specific selection rules for collisional processes.

The collisional selection rules are related with the symmetry of the molecular shape and the interaction potential of the molecules and atoms. The potential can be satisfactorily written as an expansion in Wigner D-functions. It turns out that in the interaction potential for D_2O with atoms, type A terms with even index J' of the Wigner function $D^{J'}_{DM}$ alternate with type B_1 terms with odd index [16]. There is a paper in which it was shown that the type A potential terms cause transitions with even ΔJ, and type B_1 terms cause odd ΔJ transitions, where ΔJ is the change in the total momentum quantum number of the molecular motion [17]. It was also shown that group theory provides the following selection rules: for type A potentials $-A \longleftrightarrow A$ and $B_1 \longleftrightarrow B_1$, and for type B_1 potentials, $-A \longleftrightarrow B_1$.

Combining these selection rules with the parity selection rules leads to transitions between σ^+- and σ^--levels being forbidden during collisions.

3.4. *Kinetic Model of Rotational Relaxation*

For a quantitative interpretation of the nonequilibrium distribution of D_2O molecules in an argon jet, a kinetic model of rotational relaxation was constructed [38], which was based on the following assumptions.

1) Rotational relaxation in σ^+- and σ^--systems takes place independently [40].

2) Inside each system, the probability of collisional transitions is given by the Polanyi-Woodall formula [40]

$$P_{ij} = C(2J + 1)\exp(-D\Delta E_{ij}/kT), \tag{18}$$

where C and D are constants, and ΔE_{ij} is the energy gap between levels.

The probability of reverse transitions is found from the principle of detailed equilibrium.

3) The slow collisional relaxation between σ^+- and σ^--systems is described by the relaxation equation

$$\mp dN^{\pm}/dt = (N^{\pm} - N^{\pm}_e)/\tau,$$

where $\Delta N^{\pm}$ is the total number of particles in the σ^{+}- and σ^{-}-systems (the index e refers to equilibrium conditions, and τ is the relaxation time between the σ^{+}- and σ^{-}-systems.

Assuming that the probability of transitions without changing the value of J, caused by the spherically symmetric term of the potential, should predominate over the probabilities of the remaining transitions, we introduced the weighting coefficient α into the program, which showed how much the probability of a transition with $\Delta J = 0$ was greater than the probability of a transition with $\Delta J \neq 0$ for an identical energy gap.

The system of kinetic equations for the populations of the rotational levels was solved numerically on the portion of the gas flow from the critical cross-section to the observation point for 27 of the lower levels of the *para*-system of D_2O. The parameters C, D, τ, and α were selected for the best fit to the values of $\Delta n/\Delta n_e$ using the method of least squares. The parameters were found using the method of steepest descents [41]. To accomplish this, only the first seven values of $\Delta n/\Delta n_e$ were used. Data for the transitions 3_{31}-4_{22} and 4_{31}-5_{24} were excluded for being insufficiently accurate.

Table 3 cites the experimental and calculated values of $\Delta n/\Delta n_e$ for the most well established parameters: $C_A = 1.5\cdot10^5$ $Pa^{-1}\cdot s^{-1}$, $C_{B_1} = 1.1\cdot10^5$ $Pa^{-1}\cdot s^{-1}$, $D_A = 5.8$, $D_{B_1} = 6.6$, $\tau^{-1} = 6.7\cdot10^2$ $Pa^{-1}\cdot s^{-1}$, $\alpha = 5$.

Table 3

Nonequilibrium coefficients for rotational transitions of the *ortho*-modification D_2O in the case of small D_2O concentrations

Transition	$n/\Delta n_e$		Transition	$n/\Delta n_e$	
	Experiment	Calculation		Experiment	Calculation
$1_{1}1$-2_{02}	0.34	0.54	4_{13}-4_{22}	0.9	0.91
2_{02}-2_{11}	0.65	0.60	3_{22}-4_{13}	1.9	1.41
2_{11}-2_{02}	0.41	0.39	3_{31}-4_{22}	1.47	1.87
3_{13}-3_{22}	0.35	0.34	4_{13}-5_{24}	2.1	3.95

Experimental data on collisional probabilities of rotational transitions (CPRT) for D_2O molecules are lacking in the literature. For comparison, seven polar molecules can be selected along with argon under the condition that the polar molecule has a dipole moment close in value to the dipole moment of D_2O. For example, if the CPRT for HF-Ar collisions, obtained in Ref. [42] from spectral line-widths, is used in Eq. (18), then we obtain the values $C = 0.75 \cdot 10^5$ $Pa^{-1} \cdot s^{-1}$ and $D = 5.8$, which are close to that found for D_2O-Ar collisions.

3.5. *Recovering the Distribution Function for Low Rotational Levels*

Using the measured population differences, it is possible to recover the absolute values of the populations at the lower rotational levels accurate to some arbitrary constant. To determine this constant, the sum of the populations at the levels is assumed to be equal to the sum of the populations obtained from calculations for the kinetic model with coefficients that best describe the experimentally observed absorption.

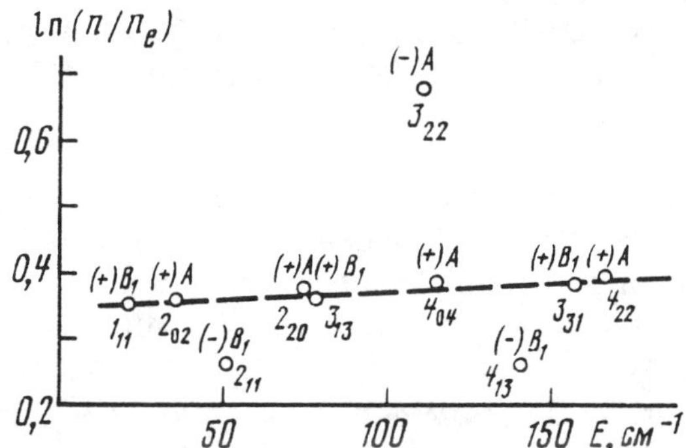

Fig. 11. Equilibrium population distribution of the lower vibrational energy levels of molecular D_2O under supersonic jet conditions. The concentration of $[D_2O] < 0.01$.

Figure 11 shows the variation of $\ln(n/n_e)$ as a function of the energy of the rotational level. Here n is the experimentally determined rotational level and n_e is the equilibrium population for a temperature of 95 K. The points on the graph correspond to the given rotational levels, the spectroscopic notation for which is shown next to the data point; also indicated is the type of wave function and its association to the σ^+- and σ^--subsystems. It is obvious that the resultant distribution can in no way be attributed to the given rotational temperature, which is exhibited in the lack of a linear variation of $\ln(n/n_e)$ with energy.

Interesting is the fact that for all levels of the σ^+-system, the experimental points lie on the line corresponding to the temperature T = 95 K. This means that the population at the lower levels of the σ^+-system can be described by a Boltzmann distribution with a temperature somewhat higher than the gas temperature. No such regularity can be found for the population of the levels of the σ^--system. The difference in the behavior of the σ^+- and σ^--systems can be explained by the density of levels in the σ^+-system being almost two times greater than in the σ^--system. The energy gap between levels 2_{11}-3_{22} is twice as great as that between levels 3_{22}-4_{13}, which, for a strong exponential dependence of the transition probability on the energy difference, leads to the relative population of levels 3_{22}-4_{13} at low temperatures being much closer to equilibrium than levels 2_{11}-3_{22}. The relative populations of levels 3_{22} and 4_{13} will also be closer to equilibrium than for the σ^+-system, since they are practically isolated, whereas there are seven closely spaced levels in the system.

When the gas temperature is lowered, molecules in the σ^+-system should accumulate at the 0_{00} level and those in the σ^--system should accumulate at the 2_{11} level. However, due to the low probability of the collisional transition 2_{11}-3_{22} at low temperatures, they accumulate at level 3_{22}.

The model, on whose basis the calculations were performed, is approximate, since the Polanyi-Woodall formula is written for the system of rotational levels of a diatomic molecule. In addition, a deviation from the exponential dependence which is built into the Polanyi-Woodall formula has recently been observed [43]. However, the model is fairly simple, permits one to get satisfactory correspondence with experiment and to qualitatively predict new results. For example, for the transitions 3_{31}-4_{22} and 4_{31}-5_{24}, which were not included in the program to find the constants, the model gives a qualitatively correct value for the total absorption.

3.6. *Case of High D_2O Concentrations*

Absorption measurements, similar to those described above, were also made for higher concentrations of D_2O in a D_2O-Ar mixture. The effect of Ar pressure on the behavior of the nonequilibrium was studied here up to pressures at which the distribution was completely equilibrium. The

Ar stagnation pressure was varied from 0.3 to 1 kPa, while the D_2O stagnation pressure was 20 Pa. It turned out that this D_2O vapor impurity was sufficient to significantly alter the form of the distribution.

In D_2O-D_2O collisions, the dipole-dipole interaction potential plays the dominant role. This potential has a symmetry different from the symmetry of the D_2O-Ar potential and, as is shown in Ref. [10], the following selection rules are valid for this potential: $|\Delta J| = 0.1$, $A \longleftrightarrow B$; and the prohibition on transition between states σ^+ and σ^- is absent.

The effect on the relaxation of the dipole-dipole interaction potential can be taken into account if a correction is introduced to the previously derived model that takes into account the corresponding selection rules. The probability of a rotational transition, related to the dipole-dipole interaction, was described by a function of the type in Eq. (18) with the parameters $\tilde{C} = 1.9 \cdot 10^5$ $Pa^{-1} \cdot s^{-1}$ and $D = 6$. The experimental results are given in Table 4 along with the calculated values.

Table 4

Nonequilibrium coefficients for the rotational transitions of D_2O in the case of high D_2O concentrations

Transition	$n/\Delta n_e$		Transition	$n/\Delta n_e$	
	Experiment	Calculation		Experiment	Calculation
1_{11}-2_{02}	0.54	0.54	4_{04}-4_{13}	1.2	1.18
2_{02}-2_{11}	0.51	0.52	4_{13}-4_{22}	1.2	1.33
2_{11}-2_{02}	0.48	0.53	3_{22}-4_{13}	1.3	1.28
3_{13}-3_{22}	0.51	0.32			

As was done above, the variation of $\ln(n/n_e)$ can be plotted as a function of the energy of the level (Fig. 12). The equilibrium population n_e corresponds to 95 K.

D_2O vapors at a concentration of 7.5% do not significantly alter the gas-dynamic properties of the supersonic jet and the adiabatic constant here varies from 1.67 to 1.64; the equilibrium gas temperature at the measurement point increases from 65 to 68 K. It is clear from Fig. 12 that the temperature of the σ^+-system, which is determined from the slope of the dashed line, remained at 84 K, i.e., very close to the equilibrium temperature. The points of the σ^--system approached the points of the σ^+-system here. Consequently, the comparatively large D_2O impurity accelerates relaxation significantly. However, as usual there is no single rotational temperature, and this can be verified in the following manner. Determining the rotational temperature from the ratio of the total absorptions for different transitions using Eq. (15), as in the previous example, its value turns out to be different for different pairs of levels.

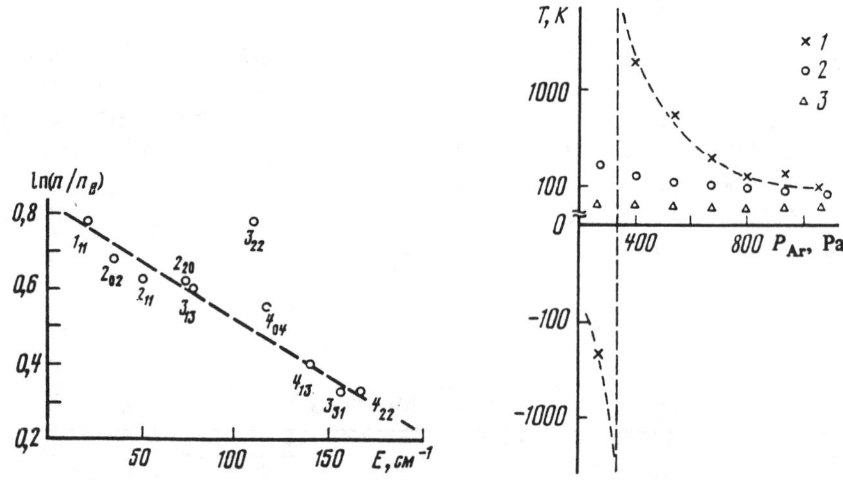

Fig. 12. Nonequilibrium distribution of the populations of the lower rotational energy levels of molecular D_2O under supersonic jet conditions. The concentration of $[D_2O] \approx 0.075$.

Fig. 13. Rotational temperature determined from the ratio of the total absorptions for different transitions. *1*) transitions 3_{13}-3_{22} and 3_{22}-4_{13}, *2*) 2_{11}-2_{02} and 4_{04}-4_{13}, *3*) 2_{02}-2_{11} and 2_{11}-2_{20}. For transitions 3_{13}-3_{22} and 3_{22}-4_{13} at P_{Ar} = 400 Pa $T \approx 2100$ K, for P_{Ar} = 267 Pa $T \approx 230$ K.

Figure 13 shows how rotational temperatures depend on argon pressure. It is clear that for transitions 3_{13}-3_{22} and 3_{22}-4_{13}, the rotational temperature, determined from Eq. (15), approaches infinity for low pressures and at 267 Pa becomes negative. For pressures of around 1 kPa for all transitions, the temperature is close to the equilibrium gas temperature. Here the rotational temperatures for all pairs of levels, lying within a single J, turn out to be close to the equilibrium temperatures for any pressure. Evidently, this indicates a large transition probability due to dipole-dipole interaction for $\Delta J = 0$.

3.7. *Measuring the Total Pressure in the Jet*

To theoretically model the gas-dynamic parameters, the flowstream is assumed to be isentropic. In order to be convinced of the validity of this assumption, we shall calculate the magnitude of the corrections attributed to the viscosity. For a viscous gas, the increase in entropy can be written as follows:

$$S = S_0 + S_1/\mathrm{Re}^* + \ldots; \tag{19}$$

where S_0 is the entropy of an isentropic flow, S_1 is the first order correction related to viscosity. The Reynolds number Re^* is calculated from the critical cross-section parameters. The Mach number can also be written in this form:

$$M = M_0 + M_1/\mathrm{Re}^* + \ldots = M_0(1 + m_1/\mathrm{Re}^* + \ldots). \tag{20}$$

As was shown in Ref. [44], the quantities S_1 and m_1 are related by the equation

$$m_1 = -\tfrac{1}{2} S_1/c_V. \tag{21}$$

The calculation of the value of m_1 is described in detail in Ref. [44] and under our conditions is $m_1 \approx -4$. Since Re^* varies from 300 to 1500, the corrections to the Mach number are on the order of 1% or lower. Thus, the calculation shows that the effect of viscosity can be neglected.

Viscosity having no significant effect on the flow can be verified experimentally by measuring the total pressure with a Pitot tube. Measuring the entropy leads to a change in the total pressure according to the equation [28]

$$\Delta S = (\kappa - 1) c_V \ln (P_{01}/P_{02}), \tag{22}$$

where P_{01} and P_{02} are the total pressures of isentropic and viscous flow, respectively. It follows from Eq. (22) that $P_{02} < P_{01}$. On the other hand, the viscosity leads to a decrease in the Mach number (Eq. (20)) and, consequently, the pressure drop across the normal discontinuity in front of the Pitot tube decreases. The latter fact weakens the pressure decrease in the Pitot tube related to the viscosity. Nevertheless, for not very high Mach numbers ($M = 3.55$ for us), the measurements of the total pressure permit one to observe the effect of viscosity fairly precisely.

So, a 10% change in the Mach number as a consequence of viscosity in the studied region would lead to a 10% change in the pressure in the Pitot tube.

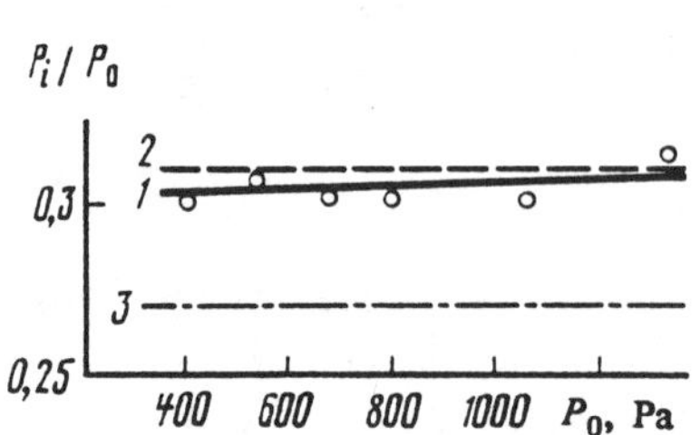

Fig. 14. Variation of the ratio of total head pressure P_i to initial pressure P_0 with P_0 for a planar supersonic Ar jet ($M = 3.55$). *1*) flow allowing for viscosity corresponding to the experimental conditions, *2*) isentropic flow, *3*) viscous flow where the Mach number is 10% below isentropic ($M = 3.2$, $T = 68$ K); the points denote experimental data.

Figure 14 shows data on the measurement of the ratio of total pressure to initial pressure for different initial Ar pressures. As is clear from the figure, the experimental points differ from the values obtained from calculations for isentropic flow by roughly 2% and lie completely along the curve calculated by Eqs. (20)-(22) which do allow for viscosity.

4. Determining the Rotational Relaxation Time of D_2O Molecules During D_2O-Ar, D_2O-He, D_2O-D_2O Collisions under Supersonic Gas Flow Conditions

The rotational relaxation rate constants obtained in the previous section can be verified by an indirect measurement of the rotational relaxation time of D_2O molecules in a supersonic flow. These measurements are also interesting because such experiments had not been done before for low gas temperatures (on the order of tens of Kelvins). As was pointed out above (see Section 2), water vapor in a supersonic flow is often found in a condensation-nonequilibrium state. Information about rotational relaxation rates of water molecules under such conditions are also lacking in the literature.

To refine our understanding of the relaxation time in a multi-level system, recall that the action of resonant radiation on the system is described with sufficient accuracy by the Bloch optical equations [45-47]. These equations are analogous to the equations of motion of a magnetic system with spin 1/2 and include relaxation times T_1 and T_2, where T_1 determines the relaxation of the population. In the case of a multi-level system, T_1 is relaxation time of the population of those levels which turn out to be in resonance with the radiation's frequency. Since the rotational spectrum of molecules with a small moment of inertia is fairly well resolved, only a pair of levels turn out to be in resonance; therefore, subsequent treatment is done in a two-level approximation.

This section describes a direct experimental determination of the rotational relaxation time of D_2O molecules at a gas temperature of 50-60 K, which is realized in a supersonic jet of inert gas. The measurement is done by the spectral line saturation technique in the submillimeter spectral region for the transitions 1_{11}-2_{02} and 2_{02}-2_{11} with frequencies of 15.6 and 13.8 cm^{-1}.

4.1. *Nonstationary Saturation in a Supersonic Gas Flow*

The investigation was done in a planar supersonic jet produced by gas effusing through a slit. The submillimeter wave beam intersected the beam parallel to the slit at some distance from it and lay in the plane of symmetry of the jet. Because of the high motion speed of the gaseous flow, the molecules are located in the zone of action of the electromagnetic field for a short time that is comparable to the rotational relaxation time. Flying through this zone, the molecules are irradiated by a pulse of the electromagnetic field, the form of which is determined by the spatial structure of the beam and the gas flow rate.

If the flight time is comparable to the relaxation time, then saturation in the moving gas is significantly different from the behavior saturation in a stationary gas. This is because, in the time for a volume of gas to move through the zone of action of the field, a stationary value of the population of the levels between which the transition is induced can not be established. On one hand, this complicates the interpretation of the experimental results, since the amount of saturation should be determined from the solution to the differential equation. On the other hand, the nonstationary nature of the process permits, as will be clear later, one to simultaneously obtain the relaxation time and the power density of the submillimeter radiation by measuring only the degree of saturation (decrease in absorption at the spectral line maximum with increasing power); the need for an independent measurement of the power density drops out here.

If the radiation is monochromatic with frequency ω, and the electric field intensity in the wave has the form

$$E = \tfrac{1}{2}\tilde{E}\exp[i(\omega t - ky)] + \text{c.c.},$$

then the population difference for a homogeneously broadened line is determined by the solution to the equation [48]

$$\frac{d(N_1 g_1^{-1} - N_2 g_2^{-1})}{dt} + \frac{N_1 g_1^{-1} - N_2 g_2^{-1}}{T_1} - \\ - \frac{N_1^e g_1^{-1} - N_2^e g_2^{-1}}{T_1} = \frac{1}{2\hbar}(g_1^{-1} + g_2^{-1})\chi''(\omega)|\widetilde{E}|^2, \tag{23}$$

where $N_{1,2}$, $N^e_{1,2}$, and $g_{1,2}$ respectively are the population, the equilibrium population, and the degeneracy of the first (second) level; T_1 is the relaxation time; and $\chi''(\omega)$ is the imaginary part of the dielectric constant. Equation (23) is valid in the presence of rapid cross-relaxation between the degenerate states of each level. In the opposite case, Eq. (23) decomposes into several equations, each of which describes saturation for some pair of upper and lower level states.

We shall introduce the coordinate system X, Y, Z (Fig. 15), where the X-axis coincides with the direction of motion of the gas flow, the Y-axis is parallel to the submillimeter wave beam, and the Z-axis is perpendicular to X and Y axes. Since the irradiating beam formed by a quasi-optical focusing system is Gaussian, the field intensity E acting on the molecules moving with the gas flow depends on time in the following manner:

$$|E(t)|^2 = |\widetilde{E}_0|^2 \exp(-z^2/R^2) \exp[-(t - t_0)^2/t_{tr}^2]. \tag{24}$$

In the above, E_0 is the intensity in the center of the beam, t_0 is instant of transit of the beam center, R is the radius of the beam, and t_{tr} is the time to transit a distance equal to R.

For a homogeneously broadened line [48]

$$\chi''(\omega) = \frac{4\pi^2}{\hbar^2}\frac{|\mu_{12}|^2}{3}(N_1 g_1^{-1} - N_2 g_2^{-1})G(\omega), \tag{25}$$

where $|\mu_{12}|^2 = \sum_{i=-I}^{I} |\mu_{ii}|^2$ is the dipole moment of the transition (since for linearly polarized radiation, applicable in our experiments, the $\Delta M = 0$ selection rule is satisfied); $G(\omega)$ is the Lorentzian line contour with central frequency Ω and line-width $1/T_2$:

$$G(\omega) = \frac{\pi^{-1} T_2^{-1}}{(\omega - \Omega)^2 + T_2^{-2}} .$$

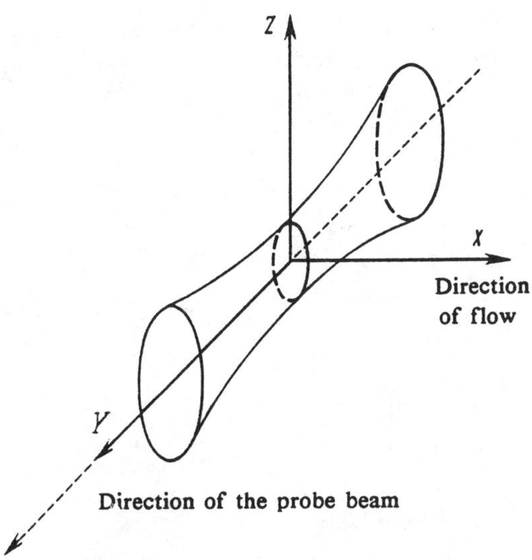

Fig 15. Submillimeter wave beam in the laboratory coordinate system.

The submillimeter radiation source that was used has a finite spectral width that is comparable with the absorption line-width. Therefore the radiation can not be considered monochromatic. The imaginary part of the dielectric constant $\overline{\chi}''(\omega)$ in this case is expressed through $\chi''(\omega)$ using a convolution over frequency:

$$\overline{\chi}''(\omega) = \int_{-\infty}^{\infty} \chi''(\omega) f(\omega - \omega') d\omega', \tag{26}$$

where $f(\omega)$ is the frequency distribution of the intensity of the BWT. If $f(\omega)$ is a Lorentzian function, then using the fact that the convolution of two Lorentzian functions is also a Lorentzian function, it results that $\overline{\chi}''(\omega)$ is expressed by Eq. (25), if $1/T_2$ in $G(\omega)$ is replaced by $1/T_2^* = 1/T_2 + 1/T_{ap}$, where $1/T_{ap}$ is the width of the distribution function $f(\omega)$. A special study of the function $f(\omega)$ was not done in our work, however, the fact that the measured experimental contour is a good approximation of a Lorentzian curve permits us to conclude that the BWT radiation spectrum is also given by a Lorentzian curve.

The Doppler broadening under our measurement conditions was 0.3 MHz, and the total contour width [Eq. (26)] was 4 Mhz. This makes it valid to regard the line as homogeneously broadened and to use Eq. (23) to find the population difference.

In the case when the frequency ω coincides with the center of the absorption line, using Eqs. (24) and (25) and denoting $N_1 g_1^{-1} - N_2 g_2^{-1}$ as n, and also making the substitution $n' = n/n_e$ and $t' = (t - t_0)/T_{tr}$, Eq. (23) can be rewritten in the form

$$\frac{dn'}{dt'} = \frac{1}{T_1'} - \frac{n'}{T_1'} + \frac{n'}{T_1'}\frac{I}{I_H}\varphi(t'), \tag{27}$$

where

$$\varphi(t') = \exp(-t'^2);$$

$$I_s = \frac{3c}{4\pi T_1' T_2^{*\prime} \hbar^{-2}(g_1^{-1} + g_2^{-1})\,|\mu_{12}|^2}. \tag{28}$$

Note that the power density at the center of the beam $I = c|\tilde{E}_0|^2/8\pi$, the saturation power density I_s, and T_1' entering into Eq. (27) form the parameter $B = I/(T_1' I_s)$ which is independent of T_1, since $I_s^{-1} \sim T_1'$. Finally, Eq. (27) has the form

$$dn'/dt' = 1/T_1' - n'/T_1' + B\varphi(t')n', \tag{29}$$

where the quantities $T_2^{*\prime}$, T_1' are expressed in units of the transit time. In deriving Eq. (29), we made use of the fact that the saturating beam is narrow and the value of n_e within it is constant.

Equation (29) as a linear homogeneous equation has a general solution that is expressed through quadrature. Since the integrals entering into the solution are not expressed in terms of elementary functions, we considered it appropriate to integrate Eq. (29) numerically by the Runge-Kutta method.

In order to determine the time T_1, it is necessary to calculate the absorption coefficient of submillimeter radiation in a supersonic flow using Eq. (27) and to compare it with experiment. The measured absorption coefficient can be represented in the form

$$\bar{\alpha} = \int_{-L}^{L} \alpha(y)dy, \tag{30}$$

where $\alpha(y)$ is the result of averaging the absorption coefficient $\alpha(x, y, z)$ over coordinates x and y (see Fig. 15) with a weight equal to the power density at a given point of the y = const cross-section.

Using Eq. (30) and assuming that $\alpha(x, y, z) \sim n(x, y, z)$, we obtain

$$\bar{\alpha} \sim \bar{n} = \int_{-L}^{L} dy \int_{-\infty}^{\infty} \int_{-\infty}^{\infty} n(x, y, z) \exp[-(x^2 + y^2)/R^2]dxdy. \tag{31}$$

When integrating over y, it must be considered that the variation of the radius R of a Gaussian beam with y is determined by the formula

$$R = R_0(1 + 3y^2/L^2)^{1/2}, \tag{32}$$

where R_0 is the radius of the beam waist and $2L$ is the path length of the submillimeter radiation in the absorbing medium. The decrease in power due to absorption as the wave propagates along the y-axis should also be taken into account.

The absorption coefficient was measured in a D_2O-inert gas mixture as a function of the inert gas pressure for radiation power densities on entering the flow I_1 and I_2 ($I_1 >> I_2$). The ratio of $I_1/I_2 = m$ was known and maintained constant. In the measurement region, the pressure P_i was varied in proportion to the stagnation pressure of the inert gas P_{i0}. The relaxation time varied in correspondence with the relation

$$1/T_1 = P_{D_2O}/\tau_{D_2O} + P_i/\tau_i, \tag{33}$$

where τ_{D_2O} is the relaxation time for D_2O-D_2O collisions per unit D_2O pressure, τ_i is a similar quantity for D_2O-inert gas collisions, and P_{D_2O} and P_i are the D_2O and inert gas pressure, respectively, in the measurement region.

Let $\overline{\alpha_1}$ and $\overline{\alpha_2}$ be the experimental absorption coefficients corresponding to the power densities I_1 and I_1/m. Since $T_1 = T_1(P_i)$, the quantity $\beta = \overline{\alpha_1}/\overline{\alpha_2}$ is also a function of P_i. On the other hand, this function can be derived from Eqs. (31) and (33) by assigning specific values to τ_i, τ_{D_2O}, and B_1, where B_1 is related with I_1 by the relation $B_1 = I_1/T_1'I_s$. During the calculation, it is necessary to consider that B_1 is proportional to T_2 and, consequently, depends on P_i:

$$B_1 = B_1^0/(1 + P_i\delta),$$

where δ is determined from the experimental measurements of the contour line-width.

The values of τ_i, τ_{D_2O}, and B_1^0, best corresponding to the experimental results, can be obtained by using the least-squares criterion. The procedure to find these parameters is described in detail in Paragraph 4.6. Note that in the stationary case ($\varphi(t')$ = const, $dn'/dt' = 0$), the desired parameters T_1 and B enter into the solution only in the form of the product T_1B and, consequently, it is impossible to determine them uniquely from the function $\beta = \beta(P_i)$.

4.2. *Description of the Apparatus and Experimental Conditions*

The measurements were done on the apparatus described in Refs. [15, 21]. Individual devices were reworked for this task. The radiation was focused by a short focal length lens (f = 2 cm) into a 1.5 mm diameter beam at the beam waist. The gases were mixed by admitting the carrier (Ar, He) and the D_2O vapors in a fore-chamber through two side-by-side openings. The jets flowing adjacently were mixed satisfactorily at a distance of a few tens of calibers due to expansion and the turbulent nature of the flow. The initial pressure of the mixture components was from 266.6 to 1066.4 Pa for Ar and He, and was always 20 Pa for D_2O. The stream passed through a slit 7.2 mm high and 20 mm in diameter. To eliminate expansion along the y-coordinate, the gas flowed inside parallel teflon walls transparent to submillimeter waves. The initial temperature of the gas mixture was room temperature in all experiments. The measurements were made at a distance of 20 mm from the slit. At the observation point, the calculated temperature was 0.19 of the initial temperature, i.e., 57 K, and the density was 0.083 of initial. The temperature and density calculations were done for an adiabatic constant $\kappa = 1.67$, since D_2O is a small impurity to the inert gas. Temperature and density nonuniformities within the beam radius did not exceed 5%.

The absorption for each component of the mixture was measured using the technique described in Ref. [21] for two power densities: I and $I/20$. In the experiments, we used linearly polarized radiation obtained using a wire polarizer. Attenuation was accomplished by introducing another polarizer into the path that was rotated by some angle. The variation of the absorption coefficient at line center with the partial pressure of He and Ar is shown in Fig. 16.

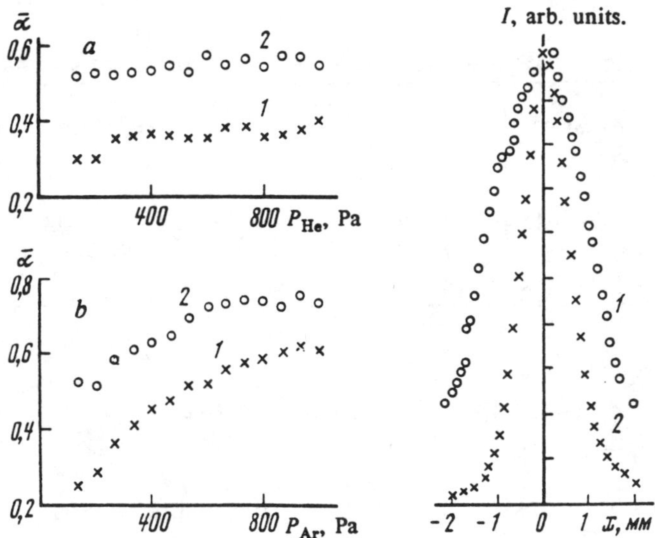

Fig. 16. Variation of the absorption coefficient at the center of the 2_{02}-2_{11} line in a mixture of D_2O-He (*a*) and D_2O-Ar (*b*) with He and Ar pressure, respectively, for power densities of *I* (*1*) and *I*/20 (2).

Fig. 17. Experimental contour of the beam of submillimeter radiation on entering the flowstream (*1*) and at the beam waist (*2*).

Several series of experiments were performed with gas-carriers Ar and He, each of which were processed by the method of least-squares. The transit time T_{tr} was determined from the flow rate of the gas, obtained by calculation, and from the profile of the beam along the *x*-coordinate.

The profile was found along the points by displacing the POAP-5 acousto-optical transducer. An aperture with a opening 0.5 mm in diameter was placed in front of the entrance window of the device. The experimental beam profile is shown in Fig. 17. The calculated flow rate of the He-D_2O mixture at the observation point was 1400 m/s, T_{tr} = 0.53 μs, and for the Ar-D_2O mixture T_{tr} = 1.67 μs. The waist of the Gaussian beam was 1.5 mm.

The results of the measurements are cited in Table 5.

Table 5

Relaxation times for different rotational transitions

Transition	Collision type	τ, μs·Pa	γ, %	Transition	Collision type	τ, μs·Pa	γ, %
2_{02}-2_{11}	D_2O-D_2O	4.2	10	1_{11}-2_{02}	D_2O-D_2O	11.2	20
	D_2O-Ar	27	4		D_2O-Ar	17	4
	D_2O-He	114	4				

4.3. *Determining the Power*

Since in the process of calculating T_1 the quantity

$$B_1^0 = I/(T_1 I_H),$$

is found, its value can be used to determine the power density. Substituting the experimental value of T_2^* and the value of T_1 calculated from the data from Table 5 into Eq. (28), we obtain $I_s = 5.8$ mW/cm^2 (T_2^* and T_1 were chosen for $P_s = 0$). From here, the power density in the center of the beam on entering the flowstream was $I = 21$ mW/cm^2.

A calculation of the transmission of the quasi-optical lens, performed for a Gaussian beam [49] for the certified values of the output power of the BWT, gives a value of 28 mW/cm^2. This correspondence can be considered satisfactory allowing for the fact that the average value of the output power in the certified data is cited over a fairly large portion of the spectrum.

4.4. *Determining the Rotational Relaxation Time of D_2O Molecules, Based on Measuring the Saturation Effect in Two Different Gas-Carriers*

Above, we showed how nonstationary saturation of a spectral transition can be realized using cw radiation. Nonstationary saturation is always more informative than stationary saturation, precisely because the relaxation times $\tau_{D_2O\text{-}D_2O}$, $\tau_{D_2O\text{-}Ar}$ can be determined from spectral measurements and the power of the radiation source alone. The method of nonstationary saturation in a gas flow (NSGF), described above, presupposes a known rate of gas flow at the observation point and likewise the variation T_1 with the composition of the mixture. This dependence [see Eq. (33)], as will be shown in Section 5, is not always valid.

Discussed in what follows is the two gas-carrier (TGC) method, based on the fact that the rate of flow of an atomic gas in a supersonic isentropic jet depends only on the mass of the atoms forming the gaseous medium. If in turn nonstationary saturation is realized for two gas-carriers with significantly different particle masses, for example helium and argon, then it is possible to separately determine the relaxation time of $\tau_{D_2O\text{-}D_2O}$, $\tau_{D_2O\text{-}Ar}$, $\tau_{D_2O\text{-}He}$, and also the power of the BWT radiation. The speed in the helium jet is 3 times higher than in the argon jet, therefore the D_2O molecules in the D_2O-He mixture will interact with the radiation for a smaller time interval than in the D_2O-Ar mixture. Therefore, saturation of the rotational transitions of D_2O molecules in the helium jet will be less than in the Argon jet.

This is expressed mathematically in the fact that a different T_{tr} will enter into the exponential of the time dependence (Eq. (24)) and Eq. (29), valid for argon, for helium in the same variables will have the form

$$dn'/dt' = 1/T_1' - n'/T_1' + B\varphi(t'\eta)n', \tag{34}$$

where η is the ratio of T_{tr} for argon to T_{tr} for helium.

In order to determine the desired variables B and T_1, we shall proceed in the following manner: we shall extrapolate the variation of $\overline{\alpha_1}$ and $\overline{\alpha_2}$ with P_g for both gases to zero carrier-gas pressure. At a carrier pressure equal to zero, relaxation will be caused only by D_2O-D_2O collisions. We shall calculate the quantities $\beta_{He}(0)$ and $\beta_{Ar}(0)$, where $\beta(0) = \alpha_1(0)/\alpha_2(0)$ is the ratio for the extrapolated values. Furthermore, by solving Eq. (29) for the determined values of T_1' and B, calculating the ratio $\overline{\alpha_1}/\overline{\alpha_2}$ using Eq. (31) and setting it equal to the experimental value, we obtain for argon

$$\beta_{Ar}(0) = F_1(T_1', B).$$

Since it is necessary to use Eq. (34) for helium, we obtain another function:

$$\beta_{He}(0) = F_2(T_1', B).$$

Solving these two equations simultaneously, we find the values of T_1' and B, where T_1' is the relaxation time for D_2O-D_2O collisions. These equations were solved graphically: the functions were plotted for both cases and the point where the two respective curves intersect was the solution.

In order to determine the interaction parameters with the carrier-gas, Eqs. (29) and (34) were solved for the previously found value of B and the T_1' was determined for a given P from equations of the type

$$\beta_{Ar}(P) = F_1^*(T_1', B), \quad \beta_{He}(P) = F_2^*(T_1', B).$$

Here, $\beta(P)$ is the ratio of $\overline{\alpha_1}/\overline{\alpha_2}$ for a given P.

The variation of $1/T_1$ for the 2_{02}-2_{11} transition with argon and helium pressure are shown in Fig. 18. As is evident from the figure, the variation is close to linear for both gases: the value of τ_g can be determined from the slope of the lines. In this manner, we found the values $\tau_{D_2O\text{-}D_2O} = 4.788$ μs·Pa, $\tau_{D_2O\text{-}Ar} = 29.925$ μs·Pa, $\tau_{D_2O\text{-}He} = 128.01$ μs·Pa, which are close to the values cited in Table 5.

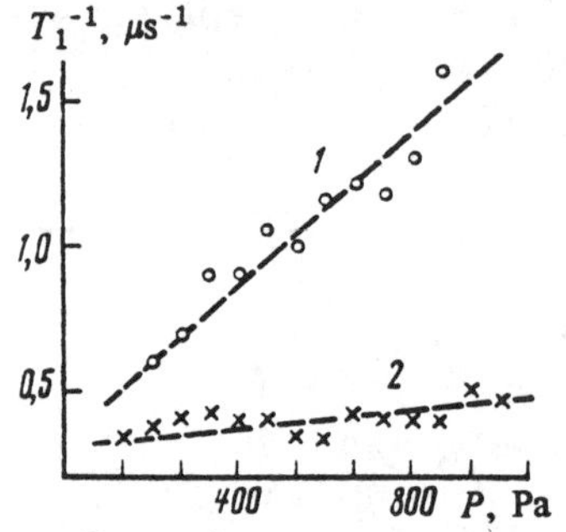

Fig. 18. Dependence of $1/T_1$ for the transition 2_{02}-2_{11} on argon (*1*) and helium (*2*) pressure.

In order to compare the relaxation times determined experimentally with those which can be calculated using the kinetic model from the previous section, we shall perform the following mathematical experiment. Solving the system of kinetic equations for the populations of the rotational levels during the flow of the supersonic jet, we shall assume that at the moment the gas flows through the plane with the coordinate = 20 mm (i.e., at the observation region) the corresponding transition is completely saturated and, consequently, the stationary value of the population difference is observed to be reestablished. By stationary value we mean only that value which was in this region in the absence of the saturating field.

The rate of recovery can determine the relaxation time T_1. For comparison, the times obtained using the NSGF and TGC methods and the calculated times are given in Table 6.

Table 6

Relaxation times values
determined by different methods at a pressure of 133.3 Pa

Transition	Collision type	T_1, μs		
		NSGF	TGC	Calc.
2_{02}-2_{11}	D_2O-D_2O	0.032	0.036	0.025
	D_2O-Ar	0.2	0.225	0.35
	D_2O-He	0.86	0.97	-
1_{11}-2_{02}	D_2O-D_2O	0.08	-	0.05
	D_2O-Ar	0.13	-	0.1

4.5. *Evaluation of the Results*

Measurements of the relaxation time T_1 for two transitions with nearby frequencies (1_{11}-2_{02} and 2_{02}-2_{11}) showed that in pure D_2O the rate of relaxation is roughly proportional to the square of the dipole moment of the transitions. This indicates that dipole interaction plays an important role during collisions of D_2O molecules.

From experiments in measuring D_2O relaxation in mixture with inert gases, it follows that D_2O-Ar collisions are much more efficient than D_2O-He collisions. This is also clear from the variation of the small-signal absorption with inert gas pressure. The absorption (under our conditions it is significantly less than equilibrium) grows with increasing initial Ar pressure and is almost constant for a similar increase in the He pressure (see Fig. 16). This result is caused by the fact that on collision with inert gas atoms, their polarizability plays an important role; for Ar it is almost 10 times greater than for He.

In order to compare the times T_1 and T_2, the absorption line-width of both pure D_2O and when in mixture with Ar and He was measured at different temperatures from 150 to 300 K, and the results were extrapolated to a temperature of 57 K. It was not possible to directly measure the line-width in the flow with sufficient accuracy due to it being so small in comparison with the instrumental width of the radiation source. It turned out that for the 2_{02}-2_{11} transition in pure D_2O, $T_1/T_2 = 4$, and on the order of 1 in a mixture with inert gases. From here it follows that in the latter case each broadening collision leads to relaxation, and resonance processes apparently play an important role in a pure gas.

4.6. *Appendix*

Let us have n experimental points y_i, which we wish to describe the function $\eta(x_1, ..., x_m)$, where x_i are the desired parameters. Using the least-squares criterion, we shall seek some function $\eta(x_1, ..., x_m)$ for which

$$\sum_{i=1}^{n} [\eta_i(x_1, ..., x_m) - y_i]^2 = \min. \tag{35}$$

Setting the partial derivatives with respect to x_j in the summation (35) equal to zero, we obtain a system of normal equations

$$\sum_{i=1}^{n} \frac{\partial \eta_i}{\partial x_j} \eta_i(x_1, ..., x_m) = \sum_{i=1}^{n} \frac{\partial \eta_i}{\partial x_j} y_i. \tag{36}$$

In the general case we obtain a system of m nonlinear equations, which by solving we can obtain the parameters x_j.

The solution to Eq. (36) can be obtained by the method of successive approximations [50].

Let $x_j^{(1)}$ be the first approximation, then

$$x_j = x_j^{(1)} + \Delta_j^{(1)}. \tag{37}$$

The function $\eta_i(x_1, ..., x_m)$ can be written in the form of the following expansion:

$$\eta_i(x_1, ..., x_m) = \eta_i(x_1^{(1)}, ..., x_m^{(1)}) + \sum_{k=1}^{m} \frac{\partial \eta_i}{\partial x_k} \Delta_k^{(1)} + ... \tag{38}$$

Substituting the values from Eq. (38) into Eq. (36), we get the system

$$b_{11}\Delta_1^{(1)} + \ldots + b_{1m}\Delta_m^{(1)} = \sum_{i=1}^{n} \frac{\partial \eta_i}{\partial x_1}(y_i - \eta_i),$$
$$\cdots\cdots\cdots\cdots\cdots\cdots\cdots\cdots$$
$$b_{m1}\Delta_1^{(1)} + \ldots + b_{mm}\Delta_m^{(1)} = \sum_{i=1}^{n} \frac{\partial \eta_i}{\partial x_m}(y_i - \eta_i), \tag{39}$$

where

$$b_{kj} = \sum_{i=1}^{n} \frac{\partial \eta_i}{\partial x_k} \frac{\partial \eta_i}{\partial x_j}. \tag{40}$$

Solving system of linear equations (39), we find the value of $\Delta_j^{(1)}$; for this value $\partial\eta_i/\partial x_j$ are calculated at the points $x_1^{(1)}$, ..., $x_m^{(1)}$. In our case, the function $\eta_i(x_1, ..., x_m)$ is not given in explicit form. It is only known that it is the solution to differential equation (27). Therefore, to calculate the partial derivative with respect to x_j, we find the solution to Eq. (27) numerically at the point $(x_1^{(1)}, ..., x_j^{(1)}, ..., x_m^{(1)})$ and at the point $(x_1^{(1)}, ..., x_j^{(1)} + \delta x_j, ..., x_m^{(1)})$, where δx_j is a small increment to the parameter x_j, and then replace the derivative $\partial\eta_i/\partial x_j$ with a difference relation.

The subsequent approximations are found in the same manner. At the lth step, the sum S_l is calculated from Eq. (35), which at first decreases and then after some number of approximations begins to oscillate around certain value. Therefore, the process of calculating the parameters stops, as long as $S_{l+1} > S_l$. It should be noted that, from a practical point of view, Eq. (37) should be used in the form

$$x_j = x_j^{(1)} + \alpha\Delta_j^{(1)}. \tag{41}$$

The quantity $\alpha \leq 1$ is chosen directly in the calculations in order to eliminate sharp discontinuities in the subsequent approximations.

5. Kinetically Isolating the Subsystem of Rotational Levels of the Heavy Water Molecules in Atomic-Molecular Collisions

The studies of D_2O molecules under rarefied supersonic argon jet conditions showed that the nonequilibrium rotational distribution of D_2O molecules can be described using a model based on the existence of kinetic equations of isolated subsystems (see Section 3). The results of the preceding section can be treated as confirmation of the numerical values of the model's parameters.

In this section, experiments are described confirming the existence of selection rules for collisional transitions in a D_2O-Ar mixture. For D_2O-Ar collisions, these rules consist in transitions being allowed only between levels belonging to a single subsystem: σ^+ or σ^-. D_2O-D_2O collisions induce transitions between levels independent of their belonging to subsystems.

The existence of two types of transitions (within and between subsystems) means that the relaxation process should have a complex character. If the concentration of D_2O is small, then on violation of equilibrium relaxation first occurs inside the subsystem with small time T_1^*, and then between subsystems with large time T_1^{**}. If equilibrium is violated with the saturation of the transition by resonant radiation, then the degree of saturation will be determined by the magnitude of T_1^* for a transition between levels belonging to one subsystem. If, on the other hand, the levels belong to different subsystems, it will be determined by the magnitude of T_1^{**}. In order for this difference to be observable, the interaction time of the gas with the radiation should be much greater than T_1.

The experiments described in Section 4 did not satisfy this condition, since the interaction time in them was on the order of T_{tr} [51]. Experiments were set up where the relaxation time was determined from the saturation effect in a motionless gas. The position of the radiation source, lens, and other details remained the same as before. This permitted us to use the value of the radiation power density determined in experiments with the jet in Section 4.

5.1. *Experimental Conditions*

The relaxation time T_1 was determined in a D_2O-Ar mixture at room temperature for the spectral transitions 1_{11}-2_{02}, 2_{02}-2_{11}, and 2_{11}-2_{20}. The levels of the first transition are referenced to the σ^+ subsystem, the two other transitions connect levels of different subsystems (2_{02} - σ^+, 2_{11} - σ^-, 2_{20} - σ^+). The frequencies of these transitions are 13.46, 15.62, and 24.8 cm^{-1}, respectively. The D_2O pressure was chosen to be 0.5 Pa, and the argon pressure was varied from 40 to 600 Pa. For high argon pressures, the rotational line of D_2O was strongly broadened, which makes it difficult to determine T_1 for two reasons: first, the accuracy in measuring the absorption index decreased; second, due to the increase in the line-width (decrease in T_2) the saturation effect decreased. Since the BWT power is finite, in addition to measurements in the lens system (as in Section 4), measurements were also done on the absorption in a circular waveguide 12 cm long and 6 mm in diameter. Due to the factor of 6 increase in the absorption length, and the increase in the relative accuracy, it became possible to work in the area of high argon pressures.

The ends of the waveguide had tapered inputs for matching with the wave traveling in free space. The time T_1 was determined using the variation of $I_s(T_1)$ (see Eq. (28)) according to the formula

$$\alpha = \frac{\alpha_0}{1 + I/I_s}, \tag{42}$$

where α, α_0 are the absorption and nonstationary absorption at the line center, I and I_s are the power density of the source and the saturation power density, respectively. The absorption in Eq. (42), as in Section 4, was averaged over the beam cross-section and over length taking into account the attenuation of the radiation. The relaxation times, determined from measurements in the waveguide and in the lens system, coincide within experimental errors for the lines 1_{11}-2_{02} and 2_{02}-2_{11} (Fig. 19, *a, b*); for the 2_{11}-2_{02} line, the measurements were done only in the waveguide (Fig. 19, *c*).

The variation of T_1 with argon pressure in the mixture has the usual character for a 1_{11}-2_{02} transition inside the subsystem: T_1^{-1} increases linearly with increasing P_{Ar} (Fig. 19, *a*). Since the saturation radiant power density I_s is inversely proportional to the relaxation time T_1, the saturation effect decreases. Therefore at high argon pressures, the measurement of the value of T_1 is difficult for this line; this can be seen clearly through the increase in the measurement errors. For $P_{Ar} > 266$ Pa, saturation at this transition is no longer observed within the measurement accuracy. For transitions between subsystems (see Fig. 19, *b, c*) the relaxation time T_1 is almost independent of P_{Ar} and at high argon pressures remains at the level of the relaxation time in pure D_2O vapors.

5.2. *Evaluation of the Results*

The experimental variations of the relaxation time T_1 with argon partial pressure shown in Fig. 19 can be compared with the calculated curves obtained via numerically modeling the experiments on saturation.

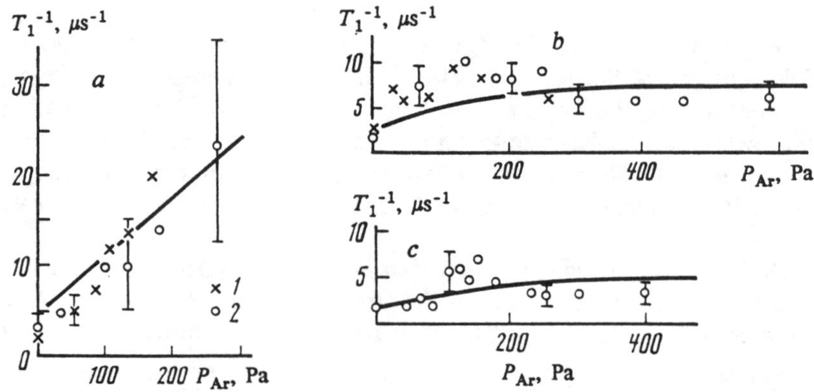

Fig. 19. Variation of $1/T_1$ with argon pressure for the transition 1_{11}-2_{02} (*a*), 2_{02}-2_{11} (*b*), and 2_{11}-2_{02} (*c*) in a D_2O-Ar mixture. *1, 2*) are the experimental points obtained from measurements for the quasi-optical lens system and for the waveguide, respectively.

For the modeling we used a system of kinetic equations with parameters that best described the nonequilibrium distribution relative to the *ortho*-modification of D_2O molecules in an argon jet (see Section 3). The effect of the radiation was accounted for by introducing the probabilities of stimulated radiative transitions for the resonant levels into the system of equations.

The value of dn_i/dt in the system was assumed to equal zero, i.e., the stationary problem was solved. Further, using the formula

$$n_1 - n_2 = \frac{n_1^e - n_2^e}{1 + I/I_s}$$

(the solution to Eq. (29) for $\varphi(t') = \text{const}$, $dn'/dt' = 0$), we determined the value of T_1.

The results of the calculation are plotted as solid lines in Fig. 19. The correspondence with experiment was satisfactory if we consider that the parameters in the system of kinetic equations were determined for the 60-120 K temperature interval.

Qualitatively, the path of the calculated curve for transitions between subsystems (see Fig. 19, *b*, *c*) can be clarified as follows: for low argon pressures, the relaxation rate will increase with increasing P_{Ar} due to the appearance of additional relaxation channels. For example, for the transition 2_{02}-2_{11}, adding argon and the appearance of D_2O-Ar collisions leads to the emergence of sequenced relaxation channels like $2_{11} \rightarrow 2_{20} \rightarrow 2_{02}$. Any chain of transitions of this type should include transitions between subsystems, the transition $2_{11} \rightarrow 2_{20}$ here. For high P_{Ar} such transitions will do little to limit the increase in the relaxation rate. In this pressure region T_1 will not depend on P_{Ar} and is determined only by the value of P_{D_2O}. We should also direct our attention to the "hump" in the vicinity of 100 Pa pressure in Fig. 18. This hump appears in connection with the fact that for these pressures, the absorption line-width becomes comparable to the spectral line-width of the BWT radiation. This leads to a decrease in the actual absorbed power and to an increase in the saturation effect.

6. Conclusion

In conclusion, we shall enumerate the possible practical applications of the results of this work.

The results obtained in this work on the basis of the relationships between rotational relaxation and the rate constants of rotational transitions for D_2O molecules can be used to model the kinetics of rotational relaxation of H_2O and D_2O molecules.

The research conducted in this work on low-temperature rarefied supersonic water vapor flowstreams showed that it is necessary to allow for rotational relaxation in them. It had been earlier thought that rotational relaxation in polar gases takes place very quickly and the rotational degrees of freedom within them are almost always equilibrium. Rotational nonequilibrium of a polar gas with a large dipole moment was established for the first time for HF molecules when studying chemical lasers.

The relationships in rotational relaxation obtained in this work can be used to produce GDL based on rotational transitions in H_2O and D_2O mixtures with atomic gases.

The application of the method of nonstationary absorption can serve to produce means for measuring the absolute power of submillimeter radiation.

REFERENCES

1. E. A. Mason, "Molecular relaxation times from thermal transpiration measurements", J. CHEM. PHYS., Vol 39 No 3 pp 522-526, 1963.

2. V. K. Annis, A. P. Malinauskas, "Temperature dependence of rotational collision numbers from thermal transpiration", J. CHEM. PHYS., Vol 54 No 11 pp 4763-4768, 1971.

3. E. V. Stupochenko, S. A. Losev, and A. I. Osipov, "Relaksatsionnye protsessy v udarnykh volnakh [Relaxation Processes in Shock Waves]", Moscow, Nauka, 1963, 484 pp.

4. E. H. Carnevale, C. Carey, and G. Larson, "Ultrasonic determination of rotational collision numbers and vibrational relaxation times of polyatomic gases at high temperatures", J. CHEM. PHYS., Vol 47 No 8 pp 2829-2835, 1967.

5. J. L. Jenkins and P. E. Wanger, "Microwave echoes in gaseous NH_3", APPL. PHYS. LETT., Vol 13 No 9 pp 308-311, 1968.

6. J. M. Levy, J. H. S. Wang, S. G. Kukolish, and J. I. Steinfeld., "Transient-nutation effect in time-resolved infra-red microwave double resonance of ammonia", PHYS. REV. LETT., Vol 29 pp 395-397, 1972.

7. K. Shimoda, "Method of double resonance in laser molecular spectroscopy", in: High-Resolution Laser Spectroscopy", New York-Berlin, Springer-Verlag, 1979.

8. V. I. Gur'ev, G. K. Vasil'ev, and O. M. Batovskii, "Measurement of the rate of rotational relaxation of HF molecules", PIS'MA ZH. EKSP. TEOR. FIZ., Vol 23 pp 256-259, 1976.

9. J. Gilbert and R. M. Vaillancourt, "A saturation effect spectrometer", PROC. IEEE, Vol 54 No 4 pp 514-520, 1966.

10. T. Oka, "Collision-induced transitions between rotational levels", ADV. ATOM. MOL. PHYS., Vol 9 pp 127-206, 1973.

11. B. N. Borzenko, N. V. Karelov, A. K. Rebrov, and R. G. Sharafutdinov, "Experimental investigation of the population of the rotational levels of molecules in a free flowstream of nitrogen", ZH. PRIKL. MEKH. TEKH. FIZ., No 5 pp 20-31, 1976.

12. R. Y. Gallaher and Y. B. Fenn, "A free jet study of rotational relaxation of molecular nitrogen from 3000-1000 K", in: "Proc. IX Symp. Rarefied Gas Dynamics", 1974, Vol 1, pp 19-32.

13. S. G. Kukolich, D. E. Gates, and J. H. S. Wang, "Rotational energy distribution in a nozzle beam", J. CHEM. PHYS., Vol 61 No 11 pp 4686-4689, 1974.

14. V. K. Konyukhov, A. M. Prokhorov, V. I. Tikhonov, and V. N. Faizulaev, "Gas-dynamic condense-laser operating on water vapor", KVANT. ELEKTRON., Vol 2 No 9 pp 2076-2077, 1975.

15. E. D. Bulatov, E. A. Vinogradov, V. K. Konyukhov, et al., "Rotational nonequilibrium of $H_2{}^{16}O$ molecules in a supersonic jet of rarefied vapor", ZH. EKSP. TEOR. FIZ., Vol 76 No 2 pp 543-550, 1979.

16. V. D. Borman, A. S. Bruev, L. A. Maksimov, and B. I. Nikolaev, "Symmetry of the interaction of molecules with rotational degrees of freedom", TEOR. MAT. FIZ., Vol 13 pp 241-247, 1972.

17. V. I. Selyakov, "Probabilities of rotational transitions of water molecules when colliding with an atom", ZH. PRIKL. MEKH. TEKH. FIZ., No 3 pp 10-19, 1980.

18. V. K. Konyukhov, "Collision selection rules in the rotational spectrum for asymmetric top type molecules", KRAT. SOOBSHCH. FIZ., No 10 pp 20-23, 1982.

19. "Kosmicheskie mazery [Cosmic Masers]", Moscow, Mir, 1984.

20. D. Dikinson, "Cosmic masers", USP. FIZ. NAUK, Vol 128 pp 345-362, 1979.

21. E. D. Bulatov, E. A. Vinogradov, N. A. Irisova, et al., "Measurement of the parameters of a supersonic flow of rarefied water vapor by the method of submillimeter spectroscopy", ZH. TEKH. FIZ., Vol 49 pp 1290-1296, 1979.

22. E. D. Bulatov, E. A. Vinogradov, N. A. Irisova, et al., "Rotational spectrum of $H_2{}^{16}O$ molecules in supersonic flows of rarefied water vapor", Preprint FIAN No 217, Moscow, 1978, 32 pp.

23. V. K. Konyukhov, "Calculation of the index of refraction for a line of the rotational spectrum of rarefied water vapor", Preprint FIAN No 195, Moscow, 1977, 36 pp.

24. C. H. Townes, "Microwave Spectroscopy", New York, Dover, 1975.

25. S. A. Clough, J. Beers, G. Klein, and L. S. Rothman, "Dipole moment of water from Stark measurement of H_2O, HDO, and D_2O", J. CHEM. PHYS., Vol 59 No 5 pp 2254-2259, 1973.

26. V. Ya. Ryadov and N. I. Furashov, "Widths and intensities of submillimeter absorption lines of the rotational spectrum of water vapor", IZV. VYSSH. UCHEBN. ZAVED. RADIOFIZ., Vol 18 pp 356-371, 1975.

27. R. T. Hall and J. M. Dowling, "Pure rotational spectrum of water vapor", J. CHEM. PHYS., Vol 47 No 7 pp 2454-2461, 1967.

28. I. V. Lebedev, "Tekhnika i pribory SVCh [Microwave Techniques and Instruments]", Moscow, Vyssh. shk., 1972, 616 pp.

29. D. D. McCracken, "A Guide to FORTRAN IV Programming", New York, Wiley, 1977.

30. K. H. Abramovich, "Prikladnaya gazovaya dinamika [Applied Gas Dynamics]", Moscow, Nauka, 1969, 824 pp.

31. M. P. Vukalovich, S. L. Puvkin, and A. A. Aleksandrov, "Tablitsy teplifizicheskikh svoistv vody i vodoyanogo para [Tables of Thermophysical Properties of Water and Water Vapor]", Moscow, Izd-vo standartov, 1969, 408 pp.

32. L. M. Davydov, "Investigation of nonequilibrium condensation in supersonic jets and nozzles", IZV. AKAD. NAUK SSSR MZhG, Vol 3 pp 66-73, 1971.

34. F. P. Boynton and A. Thomas, "Numerical computation of steady, supersonic, two-dimensional gas flow in natural coordinates", J. COMPUT. PHYS., Vol 3 pp 379-387, 1969.

35. P. A. Skovorodko, "Vrashchatel'naya relaksatsiya pri rashirenii gaza v vakuum [Rotational relaxation with a gas expanding into a vacuum]", in: "Dinamika razrezhennykh gazov [Rarefied Gas Dynamics]", Novosibirsk, ITF CO AN SSSR, 1976, pp 91-102.

36. B. F. Gordiets, A. I. Osipov, L. A. Shelepin, "Kineticheskie protsessy v gazakh i molekulyarnye lazery [Kinetic Processes in Gases and Molecular Lasers]", Moscow, Nauka, 1980, 512 pp.

37. L. D. Landau and E. M. Lifshits, "Quantum Mechanics: Nonrelativstic Theory", New York, Pergamon, 1977.

38. V. I. Tikhonov, "Nonequilibrium rotational function of molecular distribution in rarefied supersonic jet of argon", in: "XII Intern. Symp Rarefied Gas Dynamics", Book Abstr. Novosibirsk, 1982, pp 496-497.

39. S. S. Bakastov, V. K. Konyukhov, and V. I. Tikhonov, "Kinetically isolated subsystems or rotational levels of heavy water molecules with atomic-molecular collisions", PIS'MA ZH. EKSP. TEOR. FIZ., Vol 37 No 9 pp 427-429, 1983.

40. J. S. Polanyi and K. B. Woodall, "Mechanism of rotational relaxation", J. CHEM. PHYS., Vol 56 No 4 pp 1563-1572, 1972.

41. N. N. Kalitkin, "Chislennye metody [Numerical Methods]", Moscow, Nauka, 1978, 512 pp.

42. J. J. Belbruno, J. Gelfan, and H. Ratiz., "Rotational relaxation rates in HF and Ar-HF from direct inversion of pressure broadened linewidth", J. CHEM. PHYS., Vol 75 No 10 pp 4927-4933, 1981.

43. J. A. Barnes, M. Keil, R. E. Kutina, and J. S. Polanyi "Energy transfer as a function of collision energy", J. CHEM. PHYS., Vol 76 No 2 pp 913-930, 1982.

44. H. Ashkenas and F. Sherman, "The structure and utilization of supersonic free jets in low-density wind tunnels", in: "Proc. IV Symp. Rarefied Gas Dynamics", New York, Academic Press, 1966, Vol 2, pp 84-105.

45. L. Allen and J. H. Eberly, "Optical Resonance and Two-Level Atoms", New York, Wiley, 1975.

46. J. D. Macomber, "Dynamics of Spectroscopic Transitions", New York, Wiley, 1977.

47. T. Shmal'ts and U. Flaiger, "Coherent nonstationary microwave spectroscopy and Fourier transform methods", in: "Lasers and Coherent Spectroscopy", New York, Plenum 1977.

48. R. Pantel and G. Putkhov, "Osnovy kvantovoi optiki [Fundamentals of Quantum Electronics]", Moscow, Mir, 1972.

49. G. Dashan and P. Mast, "Transformation of a beam when

propagating through a system of quadratic lenses", in: "Symposium on Quasi-Optics" (M. Crowel and R. Krieger, editors), Brooklyn, New York, Polytechnic Press of the Polytechnic Institute of Brooklyn, Interscience,1964.

50. T. A. Agekyan, "Osnovy teorii oshibok dlya astronomov i fizikov [Fundamentals of Error Theory for Astronomers and Physicists]", Moscow, Nauka, 1968, 148 pp.

51. E. D. Bulatov, E. A. Vinogradov, V. K. Konyukhov, et al., "Determining the rotational relaxation time of D_2O molecules with collisions of D_2O-D_2O, D_2O-Ar, D_2O-He under supersonic gas flow conditions", Preprint FIAN No 24, Moscow, 1981, 23 pp.

EFFECT OF ROTATIONAL SELECTIVITY DURING THE HETEROGENEOUS CONDENSATION OF HEAVY WATER

V. K. Konyukhov, A. M. Prokhorov,
V. I. Tikhonov, and V. N. Faizulaev

The vapor-liquid phase transition in its initial stage is associated with the formation of dimer molecules of the condensing gas. The formation of molecules from atoms or complex molecular compounds from simpler compounds is usually controlled by the internal state of the particles making up the compound. An example of this would be the relationships in the formation of excimer molecules and spin-controllable chemical reactions. It might be suggested that the probability of dimer molecule and small cluster formation also depends on the internal state of the molecules making up the composition of the complex, such that the vapor-liquid phase transition turns out to be selective relative to the molecules' internal degrees of freedom.

In the present work, using quantitative spectroscopy of the rotational transitions of water molecules, it is demonstrated that D_2O molecules condense only from the rotational ground state. At all stages, the demonstration rests substantially on knowing the qualitative and quantitative relationships in rotational relaxation of D_2O molecules in the gaseous phase. The basic indicator of condensation is to be a reduction in the number of molecules in the vapor phase and an increase in the gas temperature due to the liberation of the latent heat of the phase transition. A brief exposition of this work was published in Ref. [1].

1. Experimental Conditions, Rotational Equilibrium of Heavy-Water Molecules, Technique for Determining Gas Temperature and Molecular Concentration

The experiments were performed on the V-2 apparatus (see Ref. [2]) with a free supersonic jet of carbon dioxide gas or nitrogen, containing a small impurity of heavy water vapor. A mixture with a constant initial heavy water vapor pressure of 0.35 Torr and a carrier-gas pressure varying within 5-40 Torr at an initial temperature of 290 K was expanded in a low pressure vacuum chamber (10^{-2} Torr) through a planar sonic nozzle (slit) with a 3 mm critical cross-section height. At a distance of 20 mm from the slit on the axis of the jet in the supersonic flow zone ($M = 4$), the total absorption coefficient was measured at the rotational transitions of D_2O molecules by the technique described in Ref. [2] (see also Refs. [3, 4]). The size of the submillimeter radiation beam in the measurement zone was 1 mm^2, the transverse size of the supersonic jet along the probe beam was 20 mm, and the calculated temperature in the measurement zone was 66 K. Industrial grade purity gases were used; the concentration of D_2O molecules in the heavy water was 99.9%.

As the gas cools in the supersonic jet, the magnitude of the total absorption coefficient α can differ from its equilibrium value α_e due to the finite rotational relaxation rate of D_2O molecules. This had been earlier investigated in detail for a sample of D_2O-Ar mixture (see Sect. 2 in Ref. [2]). In turn, the value of α_e can vary due to the condensation process, since the temperature of the flowing gas increases due to the liberation of the phase transition latent heat and D_2O molecules disappear from the gas phase, going into a condensed state, if the condensation proceeds with their participation.

As a result of the experiment, a set of curves were obtained that represent the variation of the absorption coefficient α as a function of initial carrier-gas pressure P_0 for different rotational transitions of the *ortho-* and *para-*modifications of D_2O molecules and different carrier-gases. In order to compare the variation of $\alpha(P_0)$ for different rotational transitions, the normalized total absorption coefficient $K = \alpha/\alpha_0$ was introduced. The normalization is performed relative to the equilibrium total absorption coefficient α_{e0} for a supersonic jet without condensation. If condensation does not occur, then $\alpha_e = \alpha_{e0}$ and the value of K coincides with the nonequilibrium coefficient, which was earlier used in the case of rotationally nonequilibrium flow (see Sect. 2 of Ref. [2]). If the gas flow is rotationally equilibrium, but if there is condensation, then $\alpha = \alpha_e$ and the value of $K = \alpha_e/\alpha_{e0}$ shows how the equilibrium total absorption varies due to the variation of temperature and concentration of molecules in the gaseous phase.

The normalized value of α_{e0} is determined by calculation from the equilibrium concentration of D_2O molecules at rotational levels and from the matrix element of the dipole moment of the radiative transition [5]. The temperature and density of molecules at each point of the supersonic portion of the jet are also found numerically. The value of α_{e0}, as every normalization constant, plays an additional role, such that it is not necessary to have high accuracy in determining it. When determining the parameters of the moving gas from the absorption lines of D_2O molecules, as is done below, the unnormalized function $\alpha(P_0)$ is used.

During condensation with the participation of D_2O molecules, the decrease in the absorption at frequencies corresponding to free D_2O molecules is related to the formation of stable and unstable molecular complexes. Mass-spectroscopy, which is usually used to study the initial stages of condensation in gas jets, records the presence of only the stable condensation products in the form of dimers, trimers, and larger scale molecular formations. Tracking the condensation process in terms of the original reagents making up the compound, and not the final products as is done in mass-spectrometry, provides information on the total number of D_2O molecules combined in complexes of arbitrary sizes, and thereby supplements the information from mass-spectrometry. A disadvantage of registering original reagents as opposed to registering final products consists in the need to measure the value of α very accurately in order to record the small change in it that is caused by condensation. The method of determining the concentration of water and heavy water molecules from absorption at rotational transitions has high accuracy, such that this deficiency turns out to be insignificant.

In the experiments described here, the condensation of water vapor is studied for the first time with the rotational states of the condensing molecules being resolved. Of course, it is more difficult to observe the rotational selectivity of the condensation process under conditions when the rotational nonequilibrium of D_2O molecules is inhibited by the expansion of the gas jet. The method of quantitative spectroscopy on rotational molecular transitions, which is used for the first time to study water vapor condensation, presupposes (see Refs. [2, 3]) that the gas flowstream is rotationally equilibrium, if not with respect to all rotational levels of the D_2O molecules, then in any case, for pairs of levels or groups of levels between which the spectral transitions are observed and the total absorption coefficient is measured.

It seems that conditions, under which there exists a regime of rotationally equilibrium expansion of the D_2O-CO_2 mixture on the portion of the flow from the nozzle slit to the observation point, could be established using the extensive data on rotational nonequilibrium flow of a D_2O-argon gas mixture. However, supersonic jets with carbon dioxide gas and argon as the carrier-gas have quite different parameters due to the difference in the adiabatic constants of atomic and molecular gases. The conditions of rotational equilibrium of a supersonic flowstream for all rotational transitions which are used below were established by special experiments, where molecular nitrogen replaced carbon dioxide gas as the carrier-gas. This change, on one hand, does not alter the parameters of the flow, since the adiabatic constants of carbon dioxide and nitrogen are identical due to the low initial gas temperature (T_0 = 290 K); on the other hand, molecular water condensation on nitrogen molecules does not occur for P_0 = 5-40 Torr, since a larger initial pressure or a lower initial temperature would be required for this to happen.

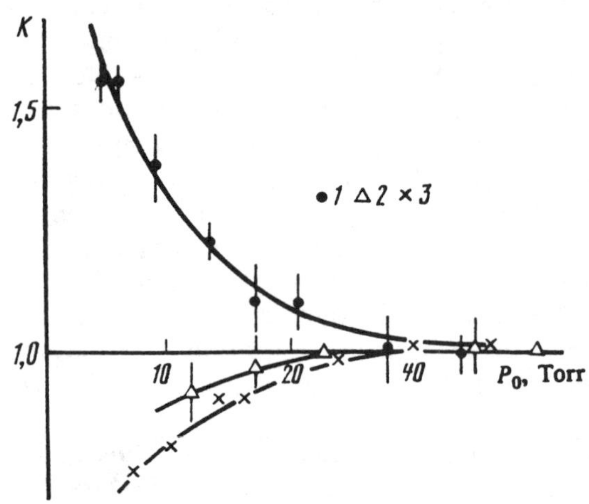

Fig. 1. Variation of K with $P_0(N_2)$ for different transitions of the *ortho*-modification of D_2O in a supersonic jet of N_2-D_2O. *1*) 4_{04}-4_{13}, *2*) 2_{02}-2_{11}, *3*) 0_{00}-1_{11}.

Figure 1 shows experimental variations of the absorption coefficient K with initial nitrogen pressure in a heavy water-nitrogen mixture for the 0_{00}-1_{11}, 2_{02}-2_{11}, and 4_{04}-4_{13} transitions of the *ortho*-modification of D_2O molecules. Nonequilibrium $K > 1$ for the upper levels and $K < 1$ for the lower levels is typical for rotationally nonequilibrium flow and is explained by the low rate of rotational relaxation in comparison with the rate of change of the temperature and density of the gas in the jet. With an increase in the initial carrier-gas pressure P_0, the relaxation rate grows directly proportional to the pressure, therefore the function $K(P_0)$ monotonically approaches unity for all spectral transitions. This form of the function $K(P_0)$ for many rotational transitions of D_2O molecules has been observed repeatedly in experiments with supersonic jets, where argon, nitrogen, and carbon dioxide gas have been used as the carrier-gas. The kinetic model of rotational relaxation, as applied to calculating the population of rotational levels of D_2O molecules (see Section 3 of

Ref. [2]), predicts a similar variation in the population difference of the rotational levels of the molecule at a given point in the supersonic jet as a function of the initial carrier-gas pressure.

It follows from Fig. 1 that D_2O molecules are in a rotationally equilibrium state in the portion of the flow from the nozzle's slit to the observation point for an initial nitrogen pressure $P_0 \geq 25$ Torr. For a flowstream with carbon dioxide gas, rotational equilibrium should be established at lower initial carrier-gas pressures, since carbon dioxide gas turns out to be a 1.5-2 times more efficient relaxant for D_2O molecules than nitrogen [6]. Calculations of the populations of rotational levels of D_2O molecules in a carbon dioxide jet, performed using the kinetic model and the rotational relaxation constants from Ref. [6], show that the function $K(P_0)$ for a jet of carbon dioxide gas is similar to the function $K(P_0)$ for a nitrogen jet with the difference that the condition for rotationally equilibrium flow being able to begin is satisfied with a carbon dioxide gas pressure $P_0 = 12$ Torr.

The condition of the D_2O molecules being in total rotational equilibrium includes not just an assumption regarding the Boltzmann nature of the distribution of molecules relative to rotational levels, but also an assumption about the equilibrium between the spin-modifications of the molecules. In the case of total statistical equilibrium, at room temperature heavy water vapor contains 1/3 molecules of the *para*-modification ($g_I = 3/9$) and 2/3 molecules of the *ortho*-modification ($g_I = 6/9$), where g_I is the nuclear statistical weight. Water vapor in the composition of the gas mixture enters the fore-chamber of the V-2 apparatus in a state of total statistical equilibrium.

It is usually assumed that the time to establish equilibrium between the spin-modifications (spin-conversion) is long, and therefore the modifications behave as independent, individually existing gases. The spin-conversion time in the gas phase is known to date only for molecular hydrogen, and is about 1.5 years [7]. For polyatomic molecules, this time is yet unknown. Recently, partial separation of the spin-modifications of CH_3F molecules has been done using the effect of light-induced drift and the spin-conversion time in the gas phase has been measured; it turned out to be 3.5 hours [8]. The process of continuous separation of water and heavy water spin-modification separation, described in Ref. [9], permitted the determination of a lower limit to the spin-conversion time of molecules (several minutes). The typical time for the gas to move from the nozzle throat to the observation point is a few microseconds, such that the assumption of the spin modifications being independent is satisfied with considerable margin. Subsequent treatment, and in particular processing the data on absorption at spectral transitions, will be done for the spin-modifications of heavy water as separately existing gases. It is further assumed that the bulk concentration of D_2O molecules of the *ortho-* and *para-*modifications are $N_o = (2/3)N$ and $N_p = (1/3)N$.

The following relationship is known between the total absorption coefficient α and the bulk concentration of the molecules as calculated for one rotational state at the lower n_1 and upper n_2 levels, between which the spectral transition is observed [5]:

$$\alpha = \int_{-\infty}^{\infty} \alpha(\omega)\, d(\omega_{21} - \omega) = \sigma_{21}\omega_{21}(n_1 - n_2); \tag{1}$$

$$\sigma_{21} = (4\pi^2 \mu^2 R)/(3\hbar c).$$

Here, ω_{21} is the angular frequency of the spectral transition, μ is the molecule's dipole moment, R is a numerical factor on the order of unity, and $\mu^2 R$ is the square of the matrix element of the radiative transition. In the case of the rotational levels having an equilibrium distribution, the molecular concentration n in a state with energy ϵ is written as

$$n = \frac{N}{Q(T)} \exp\left(-\frac{\epsilon}{T}\right), \tag{2}$$

where N is the bulk concentration of molecules of one modification, and $Q(T)$ is the statistical sum for that same modification.

From Eqs. (1) and (2) it follows that, having measured the total absorption at two transitions ($1 \rightarrow 2$, $3 \rightarrow 4$) in an equilibrium gas, it is possible to determine the gas temperature T by using the relation

$$\frac{\alpha_{21}}{\alpha_{43}} = \frac{\sigma_{21}\omega_{21}}{\sigma_{43}\,\omega_{43}}\;\frac{\exp(-\epsilon_1/T) - \exp(-\epsilon_2/T)}{\exp(-\epsilon_3/T) - \exp(-\epsilon_4/T)}. \tag{3}$$

and from Eq. (1), for the equilibrium case:

$$\alpha_{21} = \sigma_{21}\,\omega_{21} N\,[\exp(-\epsilon_1/T) - \exp(-\epsilon_2/T)]\,/Q(T), \tag{4}$$

the concentration of molecules N of one modification. Equations (3) and (4) remain valid in the case when the $Q(T)$ is understood to be the statistical sum for a group of levels with a Boltzmann distribution in which there are two pairs of levels, between which a radiative transition is observed. In this case, N is the bulk concentration of molecules at this group of levels.

In practice, the spectral transition frequencies usually turn out to be close ($\omega_{21} \approx \omega_{43}$), such that an equation can be derived from Eq. (3) which can be used to calculate the error $\Delta T/T$ in determining the temperature from the error $\Delta\alpha/\alpha$ in measuring the total absorption coefficient:

$$2 \frac{\Delta\alpha}{\alpha} = \frac{\epsilon_3 - \epsilon_1}{T} \frac{\Delta T}{T}. \tag{5}$$

From Eq. (5), it follows that the further apart the spectral transitions are spaced, the higher the accuracy in determining the temperature is. The absorption coefficient is usually measured with an error of $\Delta\alpha/\alpha$ = 3-5%, such that with $(\epsilon_3 - \epsilon_1)/T = 2.5$, the error in $\Delta T/T$ = 2-4%; for T = 60-70 K, the accuracy in determining the temperature ΔT = 1-3 K.

2. Heterogeneous Condensation of Heavy-Water Spin-Modifications

We shall examine the experimental variation of the absorption coefficient of D_2O molecules as a function of the initial carrier-gas pressure for the case of a carbon dioxide gas jet. Figure 2, *a* shows the variation of $K(P_0)$ for the transitions 2_{02}-2_{11} and 4_{13}-4_{22} of the *ortho*-modification of D_2O molecules. For small initial pressures, the curves in Fig. 2, *a* behave in correspondence with the calculation, which predicts that the function $K(P_0)$ monotonically approaches unity in the pressure region $P_0 > 12$ Torr and a constant value for the function $K(P_0)$ for higher pressures. The deviation of the experimental curve of $K(P_0)$ from unity in the pressure interval $P_0 > 12$ Torr (when the gas mixture flow is rotationally equilibrium for D_2O molecules) indicates the presence of condensation with the participation of heavy water molecules, which changes the temperature of the flowing gas and the concentration of the absorbing D_2O molecules, since in an equilibrium gas only these two factors can change the magnitude of the equilibrium total absorption coefficient α_e. For example, the fact that the value of $K = (\alpha_e/\alpha_{e0}) < 1$ in the $P_0 > 12$ pressure interval for the 2_{02}-2_{11} transition, located somewhat above the rotational ground state, indicates a decrease in the concentration of D_2O molecules in the free state, including those at levels between which the transition is observed. It also indicates an increase in the gas temperature, which causes the molecules to be shifted from levels 2_{02} and 2_{11} to higher rotational levels.

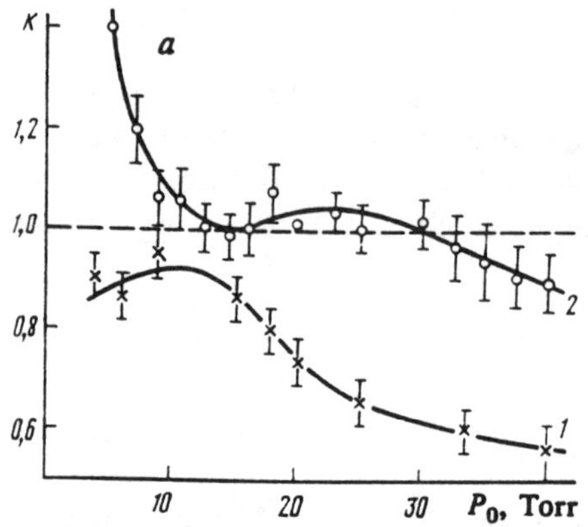

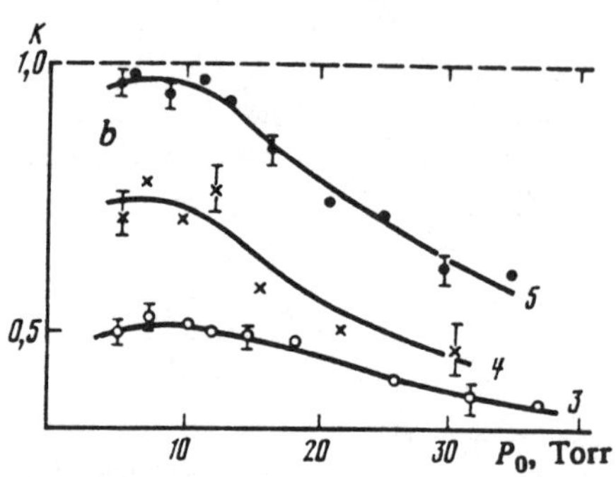

Fig. 2. Variation of K with $P_0(CO_2)$ for transitions of the *ortho*-modification of D_2O in a CO_2-D_2O supersonic jet. *1*) 2_{02}-2_{11}, *2*) 4_{13}-4_{22}, *3*) 0_{00}-1_{11}, *4*) 1_{11}-2_{02}, *5*) 2_{11}-2_{20}.

It turned out that for D_2O-CO_2 molecular vapor at a concentration of water vapor in the jet, corresponding to an initial water pressure of 0.35 Torr and an initial gas mixture temperature of 290 K, the formation of dimer molecules, their stabilization, and their subsequent growth into clusters with the liberation of the bonding energy in the form of heat begin later in relation to carrier-gas pressure than the region of rotationally equilibrium flow for D_2O molecules. This cannot be. If the region where rotationally nonequilibrium flow and condensation were to overlap, then it would turn out to be much more complex to obtain quantitative information about the condensation process.

The assumption of an equilibrium distribution of D_2O molecules over all rotational levels cannot be valid. The extreme lower rotational levels cannot be in equilibrium because of condensation. If condensation of carbon dioxide gas in the jet takes place with the participation of D_2O molecules, then we should expect molecules to vacate the extremely low levels. The lifetime of an unstable dimer due to liberated energy of rotational excitation of the molecules is less the greater this energy is, therefore, the transition to a stable form due to collisions with a third particle and the subsequent growth of the cluster occurs with a greater probability for dimers formed from weakly excited or entirely unexcited molecules.

The distribution function of D_2O molecules relative to the rotational levels of the *ortho*-modification can be plotted starting from the absorption coefficient values at the 0_{00}-1_{11}, 1_{11}-2_{02}, 2_{02}-2_{11}, 2_{11}-2_{20}, and 4_{13}-4_{22} transitions. The first four transitions relate the five lower rotational levels given in order of increasing energy. The curves of $\alpha(P_0)$

for the indicated spectral transitions in the initial pressure region P_0 = 12-40 Torr are used in this work in the case of a carbon dioxide jet with condensation (see Fig. 2). The nonequilibrium distribution function, the gas temperature T, and the number of molecules per unit volume N_0 for the *ortho*-modification in the free state are determined by a series of steps.

It turns out that the distribution function is equilibrium for a group of nine levels (the lower level is 2_{02}, the upper level is 4_{22}). This is established by comparing the values of the gas temperature T, calculated according to Eq. (3) using the absorption coefficients for the transitions 2_{02}-2_{11}, 2_{11}-2_{20}, and 4_{13}-4_{22}.

Figure 3 shows the values of $\Delta T = T - T^{0e}$ as a function of P_0 in the P_0 = 12-40 Torr pressure interval, determined with respect to the transitions 2_{02}-2_{11} and 4_{13}-4_{22}. Here, T^{0e} = 66 K is a normalization constant representing the gas temperature at the observation point and is calculated in the assumption that there is no condensation in the jet. The concentration of molecules at any pair of levels from this group in calculating one state is found from Eq. (1) from the amount of absorption at the transition between these levels (for example 2_{11}-2_{20}) and the equilibrium condition $N_2/N_1 = \exp[-(\epsilon_2 - \epsilon_1)/T]$ with the gas temperature determined above. The concentration of molecules at the remaining levels of the group is also found from the equilibrium condition with temperature T. The concentration of molecules at the lower levels 0_{00} and 1_{11}, where the distribution function is nonequilibrium, is found in correspondence with Eq. (1) from the value of α, starting from the concentration at level 2_{02}, and then for levels 1_{11} and 0_{00}. After this, the concentration of molecules in calculating one state turns out to be determined for the spectral region from the ground state to level 4_{22}. Then, calculating the degeneracy of the rotational levels, the concentration N_o of the *ortho*-molecules in the gas phase is determined.

Figure 3 also shows the variation $\Delta N_o = (N_o - N_o^{0e})/N_o^{0e}$ in the concentration of D_2O molecules of the *ortho*-modification with change in the initial carbon dioxide gas pressure in the P_0 = 12-40 Torr interval. The normalization constant N_o^{0e}, representing the bulk concentration of *ortho*-modification molecules at the observation point, is calculated in the assumption that there is condensation in the jet, the temperature of the flowing gas is increased, but that no molecules leave the gas phase.

Using the equilibrium distribution, the portion of molecules of the *ortho*-modification, which was not registered in the experiments just described, can be estimated. At levels situated above level 4_{22} for T = 60 K this share is 3% of the particles, such that the conclusion regarding the decrease in the number of molecules in the gas phase during condensation pertains to 97% of the overall number of free molecules.

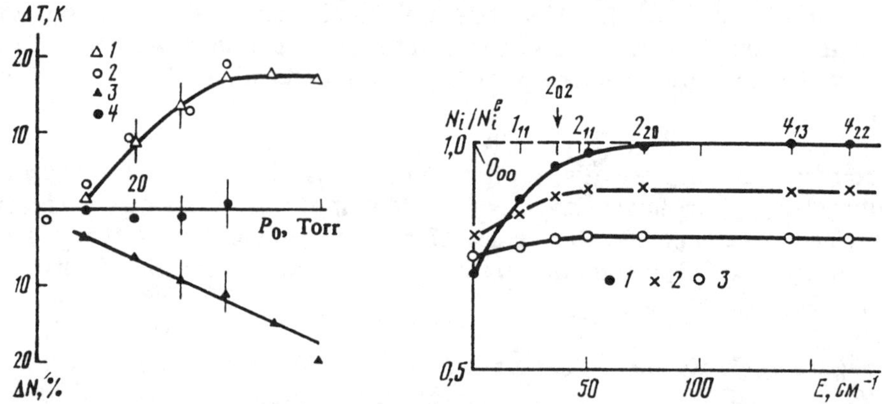

Fig. 3. Gas-dynamic parameters of D_2O molecules in a CO_2-D_2O jet as functions of the initial CO_2 pressure. *1, 2*) ΔT; *3, 4*) $(N-N^{e0})/N^{e0}$ for the *ortho*-modification (*1, 3*) and the *para*-modification (*2, 4*).

Fig. 4. Rotational distribution function for the *ortho*-modification of D_2O for different $P_0(CO_2)$. $P_0(CO_2)$: *1*) 12, *2*) 30, *3*) 40 Torr.

Figure 4 shows the unnormalized population distribution functions for the lower rotational levels of D_2O molecules for a few initial carrier-gas pressures. These curves show that with an increase in the carbon dioxide gas content in the jet, the distribution function becomes smoother, the "gap" in the first two levels becomes smaller, and the overall number of molecules in all levels simultaneously decreases.

The lower concentration of D_2O molecules at levels 0_{00} and 1_{11} than follows from the equilibrium distribution (see Fig. 4) indicates that molecules go into condensate from the lower rotational levels. Proof that D_2O molecules leave the gas phase only through the rotational ground state, while the deficit of molecules at level 1_{11} is explained by the finite relaxation rate of the transitions at the 0_{00}, is based on calculations according to the kinetic model of rotational relaxation. A negative source of particles is introduced to the system of rate equations with the relaxation constants of the D_2O-CO_2 mixture. This negative source describes molecules escaping rotational levels of different energy. The probability of escape depends exponentially on the energy of the level and the argument of the exponential is a free parameter.

We did calculations by the least-squares method for the best fit of the experimental values of the populations of levels 0_{00}, 1_{11}, 2_{02} to the calculated values for different arguments of the exponential. It turned out that the least error corresponds to a negative source of particles only in the rotational ground state with a probability of 0.1 per collision with a CO_2 molecule for a gas temperature of 60-70 K. The intensity of the particle source at excited rotational levels was negligibly small in comparison with the intensity of the source at level 0_{00}.

This is the situation with the *ortho*-modification of heavy water during heterogeneous condensation. It enables us to predict how *para*-modification molecules should behave during condensation. The rotational level with the lowest energy 1_{01} of the *para*-modification belongs to the same rotational multiplet of the asymmetric cloud as the first excited level 1_{11} of the *ortho*-modification. No noticeable migration of molecules into the condensed state from this level was observed. Similar behavior of molecules at the 1_{01} level can be expected, such that the transition of *para*-modification molecules into the condensed state is unlikely and, correspondingly, a decrease in the concentration of molecules of this modification in the gas phase should not be observed.

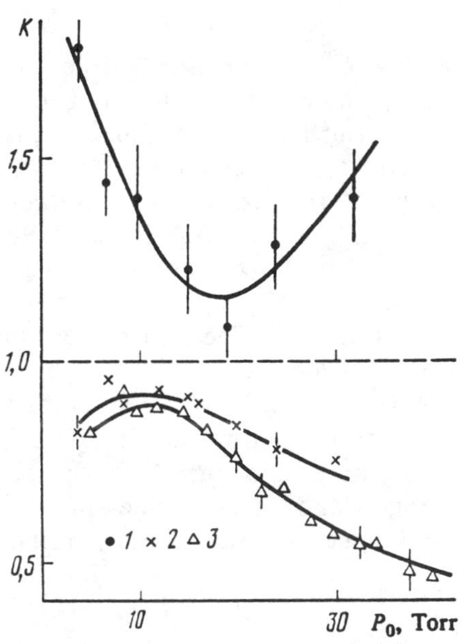

Fig. 5. Variation of K with $P_0(CO_2)$ for different transitions of the *para*-modification of D_2O in a jet of CO_2-D_2O. *1*) 5_{14}-5_{23}, *2*) 3_{03}-3_{12}, *3*) 1_{01}-2_{12}.

We shall examine the behavior of *para*-modification molecules in a carbon dioxide jet during condensation. Figure 5 shows normalized total absorption coefficients K as functions of the initial carbon dioxide gas pressure P_0 for the 1_{01}-2_{12}, 3_{03}-3_{12}, and 5_{14}-5_{23} transitions. The 1_{01}-2_{12} spectral transition starts from the same lower rotational level as the *para*-modification. It is chosen because it is convenient to monitor the change in molecular concentration in the gas phase during condensation by the amount of absorption at this transition. Two other transitions, fairly far removed along the spectrum, are suitable for determining the gas temperature by the *para*-modification molecule's absorption lines.

In the range of initial pressures $P_0 < 12$ Torr, the behavior of the curves $K(P_0)$, which were obtained under the same experimental conditions as the analogous curves for the *ortho*-modification, reflect the heavy water molecules striving for a state of rotational equilibrium in proportion to the increase in carrier-gas pressure. It is similar to the behavior of the curves of $K(P_0)$ for transitions of the *ortho*-modification with nitrogen and carbon dioxide as the carrier-gases (see Figs. 1, 2).

The curves of $\alpha(P_0)$ for transitions 1_{01}-2_{12}, 3_{03}-2_{12}, 5_{14}-5_{23} in the P_0 12-40 Torr initial pressure range were processed according to Eq. (3) with the aim of determining the gas temperature. The results showed that for each value of P_0 there are three possible pairs of transitions which, entering into Eq. (3), give matching values of the gas temperature. This means that the group of 13 rotational levels of the *para*-modification (1_{01} lower level, 5_{23} upper level) are found in a state of rotational equilibrium. An estimate of the fraction of the *para*-modification molecules which are found at the levels of this group, similar to that which was made previously for the *ortho*-modification, gives a value of 98%, such that all further evidence pertains to this portion of the free molecules of the *para*-modification.

Figure 3 shows the values of $\Delta T = T - T^{0e}$, determined from the absorption at *para*-modification transitions by the means indicated above as functions of P_0 over the P_0 = 12-40 Torr initial carbon dioxide gas pressure range. Within measurement errors, it matches the gas temperature determined by absorption at spectral transitions of *ortho*-modification molecules. The variation $\Delta N_p = (N_p - N_p^{0e})/N_p^{0e}$ in the bulk concentration of gas-phase *para*-molecules is also shown, as determined from Eq. (4) using the transitions 1_{01}-2_{12}, 3_{03}-3_{12}, 5_{14}-5_{23}, as functions of P_0 in the P_0 = 12-30 initial pressure interval. The normalization constant N_p^{0e} has the same meaning that the analogous quantity has for the *ortho*-modification. Within measurement errors $\Delta N_p = 0$, such that *para*-modification molecules remain in the gas phase and do not undergo a transition into the condensed state. Their bulk concentration during heterogeneous condensation does not change, and the change in the amount of absorption at *para*-modification spectral transitions reflects only an increase in the gas temperature.

3. Evaluation of the Results

The increase in the temperature of a gas mixture consisting of carbon dioxide gas and heavy water, and the increase in the concentration of free D_2O molecules (see Fig. 3) attests to condensation actually occurring in the supersonic jet. This can be either heterogeneous condensation of CO_2 molecules on D_2O molecules, which there are much less of than CO_2 molecules, or homogeneous CO_2-CO_2 and water-water condensation. The latter case should be excluded because of the small concentration of water molecules in the mixture's composition. Experiments with nitrogen jets in different variants show that there is no noticeable condensation of water molecules on water molecules. Homogeneous condensation of carbon dioxide gas does not happen either, as additional experiments with varying the water vapor condensation have shown. In addition to decreasing the water vapor content in the mixture, external indications of the condensation process disappear. This is expressed in the value of ΔT approaching zero, i.e., there is no increase in the temperature of the flowing gas as a low concentration of water vapor disappears. $CO_2...D_2O$ dimers, which were recently detected in a supersonic jet from their microwave spectrum [10], serve as the nuclei of heterogenous condensation of the carbon dioxide gas. The dimers have a planar T-type configuration and the calculated magnitude of the bonding energy is 25.7 kJ/mole [11], which is close to the bonding energy of water dimers.

We shall examine the molecular mechanism of rotational selectivity in the initial stages of heterogeneous condensation of water vapor. A spatial effect during the capture of molecules with an anisotropic interaction potential lies at the base of this mechanism. A water molecule as an asymmetric rotator has a dipole moment that is oriented in space only in the rotational ground state. In a gas, such molecules are directionally selective by the centers of attraction or repulsion during dipole-quadrupole interaction. In principle, capture is also possible with the participation of rotationally excited water molecules. However, the appearance of repulsion, following attraction (due to the rotation of the molecule) weakens the bond and leads to rapid disintegration of the complex. A direct cause of the of the formation of long-lived dimers with the participation of water molecules in the rotational ground state may be rotational predissociation resonance [12, 13] with the excitation of CO_2 molecules and the subsequent distribution of the energy over intermolecular vibrations of the complex.

The thermodynamically nonequilibrium state of water vapor which consists in the deviation of the molecular concentrations of *para*- and *ortho*-modifications in the gas phase from their equilibrium relationship $((N_p/N_o)_e \neq N_p/N_o)$, was judged earlier as the cause which can explain the disparity of the experimental and calculated values of the equilibrium absorption coefficients at the rotational transitions of water molecules [5]. This disagreement between theory and experiment is examined once more here, since rotationally selective condensation is indeed the source of this thermodynamically nonequilibrium water vapor.

In Refs. [14, 15], the total absorption coefficients of water vapor were measured in earth's atmosphere at room temperature at the rotational transitions 2_{02}-2_{11} of the *para*-modification and 1_{01}-1_{10} of the *ortho*-modification of molecular $H_2{}^{16}O$. For molecular H_2O, the systems of rotational levels of the *ortho*- and *para*-modifications change places in comparison with molecular D_2O, such that the rotational ground state of the H_2O molecules belongs to the *para*-modification. A calculation of the total absorption coefficient was done for transitions between the lower rotational levels of molecular H_2O, for which the rigid rotator approximation is well satisfied, and the premises and assumptions related to the quantum-mechanical and statistical portions of the treatment were analyzed [5]. The theoretical and experimental values of the absorption coefficient α/ρ normalized to the water vapor density ρ, for the transitions in the *para*- and *ortho*-systems of rotational levels are cited in Table 1. For the *ortho*-modification, the calculated and experimental values of the absorption coefficients agree within measurement errors; for the *para*-modification, the experimental value is 10% less than the calculated value, and this disparity is beyond the limits of experimental errors.

Table 1

Total absorption coefficient of water vapor at a temperature of 300 K

Spin-modification	Spectral Transition	α/ρ, cm/g	
		Theory	Experiment
Ortho	1_{01}-1_{10}	1785	1720 ± 70
Para	2_{02}-2_{11}	1185	1070 ± 40

It may be suggested that this disparity is explained by rotationally selective condensation of water vapor, which decreases the concentration of *para*-modification molecules in the gas phase, leaving the concentration of the *ortho*-modification unchanged. According to the data of Ref. [16], the relative concentration of N_2-H_2O dimers in the earth's atmosphere under normal conditions is 0.5%. If we consider the fact that dimers are formed only from water molecules of the *para*-modification, then the decrease in the bulk concentration of the *para*-modification is 2%, which agrees within an order of magnitude with the decrease in the absorption at the 2_{02}-2_{11} transition in comparison with the value calculated for total statistical equilibrium (see Table 1). For there to be a nonequilibrium state with respect to *spin*-modification of water vapor in the earth's atmosphere, it is necessary that the initial stage of condensate formation pass through nitrogen-water dimers and for the spin-conversion process of H_2O molecules to be unable to recover the disturbed equilibrium.

REFERENCES

1. V. K. Konyukhov, A. M. Prokhorov, V. I. Tikhonov, and V. N. Faizulaev, "Rotationally selective condensation of heavy water in a supersonic flowstream of carbon dioxide gas", PIS'MA ZH. EKSP. TEOR. FIZ., Vol 43 No 2 pp 65-67, 1986.

2. V. I. Tikhonov, "Rotationally nonequilibrium processes in water vapor and heavy water vapor under supersonic flow conditions", this volume.

3. E. D. Bulatov, E. A. Vinogradov, N. A. Irisova, et al., "Measurement of the parameters of supersonic flow of rarefied water vapor by the method of submillimeter spectroscopy", ZH. TEKH. FIZ., Vol 49 No 6 pp 1290-1296, 1979.

4. S. S. Bakastov, V. K. Konyukhov, and V. I. Tikhonov, "Nonequilibrium rotational distribution function for D_2 molecules in a rarefied supersonic flowstream", ZH. PRIKL. MEKH. TEKH. FIZ., No 3 pp 28-36, 1985.

5. V. K. Konyukhov, "Calculation of the index of refraction for lines of the rotational spectrum of rarefied water vapor", Preprint FIAN No 195, Moscow, 1977, 36 pp.

6. S. S. Bakastov, V. K. Konyukhov, V. I. Tikhonov, and T. L. Tikhonova, "Determining the rotational relaxation rate constant of D_2 molecules in supersonic jets of N_2 and CO_2 by the method of nonequilibrium distributions", Preprint IOFAN No 225, Moscow, 1985, 17 pp.

7. A. Farkas, "Ortovodorod, paravodorod, i tyazhelyi vodorod [Ortho-Hydrogen, Para-Hydrogen, and Heavy Hydrogen]", Moscow, ONTI, 1936, 244 pp.

8. P. L. Chapovsky, L. N. Krasnoperov, V. N. Panfilov, and V. P. Strunin, "Studies on separation and conversion of spin modifications of

CH_3F molecules", J. CHEM. PHYS., Vol 97 No 3 pp 449-455, 1985.

9. V. K. Konyukhov, V. I. Tikhonov, T. L. Tikhonova, and V. N. Faizulaev, "Distribution of spin-modifications of water and heavy water molecules", PIS'MA ZH. TEKH. FIZ., Vol 12 No 23 pp 1438-1441, 1986.

10. K. N. Peterson and W. Klemperer, "Structure and internal rotation of H_2O-CO_2, HDO-CO_2, D_2O-CO_2 Van der Walls complexes", J. CHEM. PHYS., Vol 80 No 6 pp 2439-2445, 1984.

11. B. Jonsson, G. Karlstrom, and H. Wemmerstrom, "Ab initio molecular orbital calculations on the water+CO_2 system: molecular complexes", CHEM. PHYS. LETT., Vol 30 No 1 pp 58-61, 1975.

12. S. Yu. Epifanov and V. N. Faizulaev, "Calculation of the rate constant for the formation of unstable dimers of diatomic molecules", Preprint IOFAN No 242, Moscow, 1984, 44 pp.

13. R. P. Certain, N. Moiseyev, "Weakly bound, strongly anisotropic Van der Walls molecules", J. CHEM. PHYS., Vol 89 No 14 pp 2974-2975, 1985.

14. V. Ya. Ryadov and N. I. Furashov, "Measurement of the parameters of the absorption line λ_{ij} = 398 μm or the rotational spectrum of water vapor", OPT. I SPEKTROSK., Vol 35 No 3 pp 433-438, 1973.

15. V. Ya. Ryadov and N. I. Furashov, "Widths and intensities of submillimeter absorption lines of the rotational spectrum of water vapor", IZV. VYSSH. UCHEBN. ZAVED. RADIOFIZ., Vol 18 No 3 pp 358-371, 1976.

16. J. J. Calo and J. H. Brown, "The calculation of equilibrium mole fraction of polar-polar, nonpolar-polar, and ion dimers", J. CHEM. PHYS., Vol 61 No 10 pp 3931-3941, 1974.

GROUP-THEORY ASPECT OF THE QUANTUM THEORY OF ROTATIONAL MOTION OF POLYATOMIC MOLECULES

V. K. Konyukhov

The basic thesis which is the foundation for this work consists in the fact that the systematic and comprehensive description of the rotational motion of objects, both classical and quantum, is constructed in four-dimensional space. The rotational motion of objects takes place in three-dimensional physical space, therefore the concepts related to this form of motion, for example, instantaneous rotational axis and angle of rotation, have an obvious geometrical meaning. Analysis shows that the obvious representations have a supplementary and illustrative nature, whereas the precise mathematical meaning of the quantities introduced is established in terms of four-dimensional space. This is because the fundamental role in describing the rotational motion of objects is played by a group of rotations in three-dimensional space, the set of elements of which form a sphere in four-dimensional equivalent space.

The thesis of the need for a four-dimensional description is founded on examples from the quantum mechanics of rotating objects. They are devoted to the quantum theory of rotation of molecules, such as: determining the parity of the rotational wave function of polyatomic molecules, the relationship between the spatial shape of the molecules and the form of the intermolecular interaction potential, and the commutation relations in a rotating coordinate system. The solution to these problems, which is given below, demonstrates the value of the group-theory approach.

In the quantum theory of rotational motion of molecules, two types of group representations for rotations in three-dimensional space are used. The regular representation in the function space assigned to the group will be used when we are concerned with polyatomic molecules of the symmetric and asymmetric rotator types, and quasi-regular representation in the function space given on a sphere in three-dimensional space will be used when were are concerned with diatomic or linear molecules. An example of water molecules scattering at an atom or a diatomic molecule is examined in this work, when the quasi-regular representation should be used for a polyatomic molecule. This approach enables us to establish new collisional selection rules for the rotational transitions of water molecules.

1. Parity of the Rotational Wave Function of Polyatomic Molecules

The operation of spatial inversion is determined in three-dimensional physical space F_3 as a mutually single-valued reflection of a set of vectors onto themselves ($\mathbf{x} \rightarrow -\mathbf{x}$), where the vector $\mathbf{x}$ is juxtaposed to its inverse ($-\mathbf{x}$) relative to the operation of vector summation. The problem consists in the transfer of this operation to H_4 quaternion space [1, 2]. The rotational wave function of a polyatomic molecule (D-function in the case of spherical or symmetric rotators or a linear combination of D-functions in the case of an asymmetric rotator) is defined on a sphere of unit radius in H_4 space, therefore its behavior on inversion can be calculated only after the inversion operation is defined on a three-dimensional sphere. For diatomic and linear molecules, the problem of transferring the inversion operation does not exist, since the rotational wave functions of these molecules (spherical or Y-functions) are defined on a sphere in three-dimensional space.

There is always an isomorphic correspondence between the physical space F_3 and the subspace of imaginary quaternions H_3. Establishing an isomorphic reflection of F_3 onto H_3 is necessary to describe rotations in F_3 in the language of quaternions or matrices. In the physical literature, this step, as a rule, is not fixed; it is regarded as obvious that each rotation in F_3 corresponds to some matrix in H_4. However, only after establishing the correspondence between the spaces does it turns out to be possible to answer the question of how the axis of quantizing the angular momentum is oriented in F_3.

Isomorphic reflection of F_3 in H_3 is established in the correspondence of the basis sets $(\mathbf{e}_1, \mathbf{e}_2, \mathbf{e}_3) \approx (\mathbf{i}, \mathbf{j}, \mathbf{k})$. In H_4 space, there is a separation of the coordinate systems $(1; \mathbf{i}, \mathbf{j}, \mathbf{k})$, where $\mathbf{i}, \mathbf{j}, \mathbf{k}$ are the quaternion unit vectors; in F_3 space the $(\mathbf{e}_1, \mathbf{e}_2, \mathbf{e}_3)$-coordinate system has the correct orientation, and is arbitrary in remaining respects. The latter requirement is related to the correspondence between the vector multiplier in F_3 and the multiplier of the quaternion units. The arbitrariness in choosing the coordinate system in F_3 reflects the isotropy of physical space and, in particular, the arbitrariness of the direction of the quantization axis of the angular momentum. In H_4 space, the direction of the quantization axis is always given by the quaternion $\mathbf{k}$.

The inversion transformation in F_3 corresponds to the quaternion conjugation operation in H_4, which is established from the correspondence of the basis vectors:

$$(-\mathbf{e}_1, -\mathbf{e}_2, -\mathbf{e}_3) \approx (-\mathrm{i}, -\mathrm{j}, -\mathrm{k}),$$
$$\mathbf{x} = x_0 + x_1 \mathrm{i} + x_2 \mathrm{j} + x_3 \mathrm{k},$$
$$\overline{\mathbf{x}} = x_0 - x_1 \mathrm{i} - x_2 \mathrm{j} - x_3 \mathrm{k}, \qquad \mathbf{x}, \overline{\mathbf{x}} \in H_4;$$

where $\overline{\mathbf{x}}$ is the conjugated quaternion.

The next step consists in establishing what the conjugation operation itself represents for the set of elements of a three-dimensional sphere. The set of quaternions, equal to unity in modulus, forms a group G by multiplication, that is isomorphic to the matrix group $SU(2)$. The correspondence between first order quaternion and complex matrices is established using the matrix realization of quaternion units:

$$1 \longleftrightarrow \begin{Vmatrix} 1 & 0 \\ 0 & 1 \end{Vmatrix}, \quad \mathrm{i} \longleftrightarrow \begin{Vmatrix} 0 & i \\ i & 0 \end{Vmatrix}, \quad \mathrm{j} \longleftrightarrow \begin{Vmatrix} 0 & -1 \\ 1 & 0 \end{Vmatrix}, \quad \mathrm{k} \longleftrightarrow \begin{Vmatrix} i & 0 \\ 0 & -i \end{Vmatrix},$$

$$\mathbf{x} = x_0 + x_1 \mathrm{i} + x_2 \mathrm{j} + x_3 \mathrm{k} \longleftrightarrow \begin{Vmatrix} x_0 + ix_3 & -x_2 + ix_1 \\ x_2 + ix_1 & x_0 - ix_3 \end{Vmatrix} \equiv u; \tag{1}$$

$$\mathbf{x} \in G, \quad |\mathbf{x}| = 1, \quad u \in SU(2), \quad \mathrm{Det}\, u = 1.$$

From Eq. (1) it follows that the conjugate quaternion is associated with the inverse matrix

$$\overline{\mathbf{x}} = x_0 - x_1 \mathrm{i} - x_2 \mathrm{j} - x_3 \mathrm{k} \longleftrightarrow \begin{Vmatrix} x_0 - ix_3 & x_2 - ix_1 \\ -x_2 - ix_1 & x_0 + ix_3 \end{Vmatrix} \equiv u^{-1}.$$

Thus, the spatial inversion operation in physical F_3 space corresponds to a mutually single-valued reflection of a three-dimensional sphere onto itself, for which the matrix of the first rotation is converted into a matrix of the inverse rotation.

It now is necessary to determine how functions, given on a sphere of unit radius in H_4, transform after reflection $u \to u^{-1}$. In the theory of polyatomic molecules' rotation, Wigner D-functions $D^{\mathrm{i}}_{\mathrm{mn}}(\varphi, \vartheta, \psi)$ are used as the wave functions or basis functions, which represent the matrix elements of an irreducible unitary representation of dimension $2l + 1$, written as functions of the angular variables φ, ϑ, ψ parameterizing the set of elements of group $SU(2)$. From the theory of group $SU(2)$ representations it is known that the matrix elements $T^{\mathrm{i}}_{\mathrm{mn}}(u)$ of the representation (after replacing the matrix argument u by u^{-1}) transform as [3]:

$$T^l_{mn}(u^{-1}) = \overline{T^l_{nm}(u)},$$

The matrix of the representation is transposed, and its elements are replaced by their complex conjugates. (The bar denotes complex conjugation.) The D-functions are transformed in the same fashion:

$$D^l_{mn}(-\psi, -\vartheta, -\varphi) = \overline{D^l_{nm}(\varphi, \vartheta, \psi)}. \tag{2}$$

The angles (φ, ϑ, ψ) define the matrix u, and the angles ($-\varphi$, $-\vartheta$, $-\psi$) correspond to the matrix u^{-1}.

The inversion group $\{E, I\}$ has two one-dimensional irreducible representations: {1, 1} and {1, -1}. From Eq. (2) it follows that on inversion in F_3, the D-functions are not transformed by any single irreducible representation, therefore the concept of parity for D-functions does not exist. This conclusion follows from Wigner's theorem, if it is applicable to the rotational Hamiltonian of polyatomic molecules.

According to this theorem, the eigen functions of the energy operator belonging to a single eigen value are transformed by an irreducible representation of that group, relative to which the energy operator is invariant [4]. Let us assume that the D-function has parity, i.e., under the action of the inversion operator it is transformed into one of two irreducible representations. Since D-functions or their linear combinations are eigen functions of the Hamiltonian, then according to the conditions of the theorem, the rotational Hamiltonian is transformed into an identical representation of the inversion group. It is shown below that the rotational Hamiltonian of a polyatomic molecule is not invariant to inversion in F_3, consequently, the assumption of parity of the D-function is untrue as well.

The rotational Hamiltonian of a polyatomic molecule is expressed in terms of the infinitesimal operators of either right A_k or left B_k displacement (k = 1, 2, 3). The operators A_k, B_k and their relationship to the angular momentum operators in different coordinate systems are discussed in detail in Sect. 4. On $u \to u^{-1}$ reflection, the A_k operators become B_k operators and vice versa. One can be convinced of this by writing the operators in terms of the angles φ, ϑ, ψ and making the substitution (φ, ϑ, ψ) $\to$ ($-\psi$, $-\vartheta$, $-\varphi$) (see Eqs. (5) and (6), below). Since the Hamiltonian is written asymmetrically in relation to the set of six operators A_k, B_k (see Eq. (4), below), then on $u \to u^{-1}$ transformation, it does not turn onto itself and, therefore, is not invariant to inversion transformation in F_3.

The asymmetry of the Hamiltonian to exchanging forward rotation with backward rotation is reflected in the terminology taken in molecular theory. The operator A_3 corresponds to the projection of the angular momentum onto the molecular coordinate system, and is not conserved in time for an asymmetric rotator; conversely, the B_3 operator corresponds to the projection onto the laboratory system of coordinates, and it is an integral of motion.

The rotational Hamiltonian of a diatomic or linear molecule, conversely, is invariant to the inversion operation in F_3, and its eigen functions (spherical functions) are transformed with respect to a continuous representation of the group inversion. This proof is easily verified. The infinitesimal operators of the quasi-regular representation in the space of homogeneous harmonic polynomials [3]

$$A_{jk} = x_k \frac{\partial}{\partial x_j} - x_j \frac{\partial}{\partial x_k}, \qquad j, k = 1, 2, 3, \quad x_j, x_k \in R_3,$$

corresponds to the molecule's angular momentum operators and the orbital momentum operators. They are invariant to the transformation x → -x. The moments of inertia of the body are likewise invariant to the inversion transformation. Spherical functions [3]

$$Y_{lm}(x_1, x_2, x_3) = A^l_m\, r^l C^{1/2+m}_{l-m}(x_3/r)\,[(x_2 \pm i x_1)/r]^m,$$

where A^l_m is a normalization factor, $r = (x_1^2 + x_2^2 + x_3^2)^{\frac{1}{2}}$, $C_p^{\frac{1}{2}+m}(-t) = (-1)^p C^{H+m}(t)$ is Gegenbauer polynomial ($p = l-m$), are transformed on inversion as

$$Y_{lm}(-x_1, -x_2, -x_3) = (-1)^l\, Y_{lm}(x_1, x_2, x_3).$$

In the physical literature on the theory of angular momentum, the operators A_{jk} and the spherical functions are expressed in terms of the angles (ϑ, φ) of the spherical coordinate system in F_3. A distinction should be made between the Euler angles (φ, ϑ, ψ), by which the elements of group $SU(2)$ are parameterized, and the spherical coordinates. They are frequently denoted by the same letters, but their meaning is different. This is clear, for example, because the angular variables (ϑ, φ) are transformed on inversion as $(\vartheta, \varphi) \to (\pi - \vartheta, \varphi + \pi)$, where as the Euler angles are transformed as $(\varphi, \vartheta, \psi) \to (-\psi, -\vartheta, -\varphi)$.

The concept of parity can be retained for the rotational wave function when going from a diatomic molecule to a polyatomic molecule, if the polyatomic molecule's Hamiltonian is written using the operators A_{jk}

or, more obviously, the orbital momentum operators L_k, $k = 1, 2, 3$. For example, the Hamiltonian of an asymmetric rotator in this case looks like

$$\mathcal{H} = (\hbar^2/2)(L_1^2/I_a + L_2^2/I_b + L_3^2/I_c), \tag{3}$$

where I_a, I_b, I_c are the principle moments of inertia of the rotator.

Going from basis D-functions to basis Y-functions changes neither the energy of the rotational state nor the expansion coefficients of the state's wave function in the basis functions, since the corresponding matrix elements of the operators L_k and J_k, $k = 1, 2, 3$, have identical values:

$$\langle l, m \mid L_k \mid l', m' \rangle = \langle l, m, n \mid J_k \mid l', m', n \rangle.$$

The second magnetic quantum number n is arbitrary, but constant.

In this approach, the concept of parity of the rotational state is related to the fact that one completely defined space of spherical functions is selected from among the several functional spaces where the group representations $SU(2)$ and $SO(3)$ are realized. From the physical point of view, writing the rotational Hamiltonian of polyatomic molecules in terms of the orbital momentum operator does not contradict the idea that the orbital angular momentum is directly related to the rotational motion of the body in three-dimensional physical space.

The traditional description of the rotational motion of polyatomic molecules using D-functions has a good experimental basis. Examined on this basis are: the interaction of molecules with an electromagnetic field, radiative selection rules, the Stark and Zeeman effects, and the heat capacity of a molecular gas. The question of how the Y-description of the rotational motion of polyatomic molecules can find application for what physical phenomena remains open. In Sect. 3, however, an example of rotational selection rules in molecular-atomic collisions is given, where basis Y-functions are used for molecular water.

The question of rotational wave function parity arises in connection with the classification of the total wave function of the molecule in relation to the spatial inversion operation [5]. In the case of nonlinear polyatomic molecules, there is a problem with defining the concept of parity of a rotational wave function. In Refs. [5-7], it was solved by replacing the inversion operation with some substitution of angular variables or transposition of the nucleus in the molecule. Replacing inversion with transposition is regarded as unsatisfactory because the invariance of the Hamiltonian of the molecule relative to the group of angular transpositions is already used to classify the state of symmetric and asymmetric rotators, therefore, new quantum numbers do not arise with the introduction of inversion in this fashion. Replacing inversion with a rearrangement of the nucleus is unsatisfactory, since additional knowledge of molecular structure is required in comparison with that contained in the moments of inertia. The values of the principle moments of inertia are the only factor of the structure of the molecule which enter into the model of rigid symmetric or asymmetric rotators.

2. Molecular Shape and the Intermolecular Interaction Potential

The representation of the spatial form of a polyatomic molecules is widely and successfully used when classifying the vibrational levels and vibrational functions of molecules. In describing rotational motion, we consider the symmetry properties of geometric figures that form the nuclei of the molecules when they are in a state of equilibrium. Using ideas of the shape of a molecule to classify rotational states and intermolecular interaction potentials is laden with difficulties, the cause of which consists in the fact that the quantum description of vibrational and rotational motions is constructed in different functional spaces.

The point group K of the molecule's symmetry acts as a group of transformations in F_3 physical space, reflecting the shape of the nuclear core of the molecules onto itself. Vectors for displacing the nucleus away from its equilibrium position and displacement operators are assigned in this same vector space. The vectors in F_3 space are the argument in the vibrational wave functions; therefore, difficulties do not arise when defining the transformation properties of the displacement operators and the functions under the action of group K operations.

In the quantum theory of rotational motion of polyatomic molecules, the function space of Wigner D-functions $D^{l}_{mn}(\varphi, \vartheta, \psi)$ is used. Given in this same space are the angular momentum operators entering into the rotational Hamiltonian as differential operators that act on functions of angles φ, ϑ, ψ. The set of points $(\varphi, \vartheta, \psi)$ form a sphere S_3 of unit radius in four-dimensional equivalent H_4 space. The point group K, being an F_3 transformation group, does not operate on the sphere S_3. This is also thee main difficulty in transferring the notion of molecular shape from vibrational motion to rotational.

The difficulty can be avoided if on sphere S_3 we construct a group of transformations K', isomorphic to group K, and further construct their classification on the basis of this new group, assuming that it carries information about the shape of the molecule. It turns out that in the course of constructing the groups, isomorphic to the point groups of the molecules, on going from group K to K' almost all knowledge of a specific molecule is lost. Another conclusion consists in creating a variant of the theory of rotational motion of polyatomic molecules, when the angular momentum operators and wave functions are defined on sphere S_2 in F_3 space.

We shall evaluate this variant in detail with the construction of an isomorphic group. Among the point groups which correspond to different shaped molecules, there are isomorphic groups between the groups themselves, for example, C_{2h}, C_{2v}, D_2, such that it and the same group K' will correspond not to a class of molecules of identical symmetry, but to a set of classes of molecules with different spatial symmetry. The isomorphic reflection of the group K onto group K' can be accomplished in several ways (these means exist as long as automorphisms of group K exist), such that there is no strict correspondence between the group K operations and the elements of group K'. This depends on the arbitrariness in constructing the reflection of K onto K'. For example, some nonunique element $c \in K'$ can correspond to a rotation of C_2, which enters into all three groups, or for another construction of isomorphic reflection, corresponds to an inversion from group C_{2h}, a reflection from group C_{2v}, and rotation from group D_2. From what has just be said, it follows that allowing for the shape of the molecule by constructing an isomorphic group representation is of little value.

We shall examine the group of invariance of the Hamiltonian of an asymmetric rotator known in the literature [5]. Since it is isomorphic to the point group D_2, then its elements are related by rotations through an angle of π relative to the axis of the molecular coordinate system, i.e., with transformations in physical space. The angular momentum operators, which enter into the Hamiltonian, except for a numerical multiplier, are infinitesimal operators A_k, $k = 1, 2, 3$, of the first regular representation [8]. Group D of the invariance of the Hamiltonian

$$\mathcal{H} = A_1^2 + A_2^2 + A_3^2 \tag{4}$$

is a group of substitutions of the angular variables φ, ϑ, ψ through which the operators A_k are expressed. The substitutions can be written in the following form:

$$(\varphi, \vartheta, \psi) \to (\varphi, \vartheta, \psi), \qquad (\varphi, \vartheta, \psi) \to (\varphi, \vartheta, \psi \pm \pi),$$

$$(\varphi, \vartheta, \psi) \to (\varphi, \vartheta + \pi, -\psi), \quad (\varphi, \vartheta, \psi) \to (\varphi, \vartheta + \pi, -\psi \pm \pi).$$

For any nonidentical substitution, the two A_k operators are multiplied by (-1), such that the Hamiltonian (4) turns onto itself and the commutation relations between operators remain unchanged. The cited form of the substitution is not unique; two other forms exist [9]. In this case it is only important that the angle substitution fall outside the limits of the range of the angular variables

$$0 \leqslant \varphi < 2\pi, \quad 0 < \vartheta < \pi, \quad 0 \leqslant \psi < 2\pi$$

on group $SO(3)$ [3]. This means that the A_k operators and, correspondingly, the D-functions, are assumed given on the entire number line, more accurately, as the direct product of three number lines. To transfer the geometric representations from F_3 to this space seems complicated, therefore the classification of rotational states of an asymmetric rotator with the Hamiltonian of Eq. (4) via rotation in F_3 space should be taken as being illustrative.

A discussion of the theory of the rotational motion of polyatomic molecules using orbital momentum was begun in the preceding section. We shall establish a group of invariance of the Hamiltonian and the classification of rotational states on that basis.

Table 1

Irreducible representations of the point group D_{2h}

	E	$C_2^{(x)}$	$C_2^{(y)}$	$C_2^{(z)}$	I	$\sigma^{(yz)}$	$\sigma^{(xz)}$	$\sigma^{(xz)}$
A	1	1	1	1	1	1	1	1
B_1	1	-1	-1	1	1	-1	-1	1
B_2	1	-1	1	-1	1	-1	1	-1
B_3	1	1	-1	-1	1	1	-1	-1
$\tilde{A}$	1	1	1	1	-1	-1	-1	-1
$\tilde{B}_1$	1	-1	-1	1	-1	1	1	-1
$\tilde{B}_2$	1	-1	1	-1	-1	1	-1	1
$\tilde{B}_3$	1	1	-1	-1	-1	-1	1	1

Note: $C_2^{(\;)}$ is rotation by an angle of π, $\sigma^{(\;)}$ is reflection in the plane (the symbols of the corresponding axes of rotation and the reflection planes are given in the parenthesis).

The invariance of the Hamiltonian **H** of a rigid asymmetric rotator, written in terms of the orbital momentum operators (Eq. (3)), relative to the inversion operation was established in Sect. 1. It is not difficult to establish the invariance of the operator **H** relative to rotations of the point group D_2. For each rotation $C_2 \in D_2$ two cartesian coordinates are multiplied by (-1), such that two operators L_k, $k = 1, 2, 3$, change sign. The total group of invariance of the Hamiltonian of Eq. (3) is the point group D_{2h}.

Group D_{2h} has eight irreducible one-dimensional representations that are cited in Table 1. We shall denote four of them as A, B_1, B_2, B_3, as the group D_2 representations, their inversion operations are even, and the four others we shall denote as $\tilde{A}$, $\tilde{B}_1$, $\tilde{B}_2$, $\tilde{B}_3$, their inversion operations are odd. The C_{2v} subgroup of group D_{2h} consists of the following transformations: $C_{2v} = \{E, C_2^{(z)}, \sigma^{(yz)}, \sigma^{(xz)}\}$.

The position on the energy scale and the number of rotational levels in going from the Hamiltonian of Eq. (4) to that of Eq. (3) remained the same, but the classification of the rotational states became more detailed. While two quantum numbers with values (+1; -1) were required for classification in terms of irreducible group D representations, there are now three such numbers, one of which is the parity of the quantum number J.

3. Collisional Selection Rules

The hypothesis on which the explanation for the collisional selection rules of water molecules is based consists in the fact that there are two forms of the quantum-mechanical description of the rotational motion of polyatomic molecules: in terms of D-functions (D-description) and in terms of Y-functions (Y-description). Each of these forms is employed as a function of which particle interacts with the water molecule during collision. If an atom, diatomic molecule, or linear rotator molecule acts as the collision partner, then Y-functions are used; if a polyatomic molecule of the symmetric or asymmetric rotator variety, then D- and Y-functions. The assignment of the forms is based on the following considerations.

The rotational motion of water molecules can be perturbed either by the relative coordinate motion of the colliding particles (RL-interaction), or by the rotational motion of the collision partner (RR-interaction). The angular part of the coordinate motion and the rotation of the linear rotator are coupled with the angular momentum and their wave functions are given in F_3 space. In this space, the rotational function of molecular water should be given such that its rotation could interact with the orbital motion and rotation of the rotator, but the wave functions interfere, and therefore the Y-description is employed. If the water molecule collides with another polyatomic molecule, then to describe the interaction of its rotation with the coordinate motion, the previous rotational states of the water molecules are written in terms of Y-functions. To describe the mutual effect of rotational motion of the molecules, the rotational states of both molecules should be written in terms of the D-functions. The RR-interaction potential can be dipole-dipole or, in general, multipole.

In the case of the D-description of RR-interaction of polyatomic molecules, the collisional selection rules of molecular water coincide with the radiative selection rules, since they are both related to the interaction of a molecular dipole moment with an electromagnetic field. In classifying the rotational states of water molecules in terms of the irreducible representations A, B_1, B_2, B_3 of group D, and when choosing the interaction potential of the dipole moment with the electromagnetic radiation in such a form that it is transformed in B_1 representation, the radiative transitions turn out to be resolved between states $A \rightleftarrows B_1$, $B_2 \rightleftarrows B_3$. The total system of rotational levels is divided into two subsystems (A, B_1), (B_2, B_3), between which radiative and collisional transitions are forbidden.

In the case of the Y-description of the rotational states, when determining the collisional selection rules it is necessary to use the concept of the spatial shape of the water molecule. It reduces to the fact that the RL- and RR-interaction potentials do not change after transformations of the point group C_{2v} of the symmetry of the water molecule, since these transformations of F_3 space bring the molecule back onto itself. Since the group C_{2v} molecular shapes are a subgroup of D_{2h} of the invariance of the rotational Hamiltonian of the molecule, it is possible to indicate the form of the RL- and RR-potentials, which are allowed by the shape of the molecule, in terms of irreducible D_{2h} group representations. Those potentials are permitted which are multiplied by unity when performing the transformations introduced to group C_{2v}, i.e., remain unchanged. According to the table of irreducible group C_{2h} representations for molecules with C_{2v} symmetry, the potentials A and $\tilde{B}_1$ are allowed. The total system of rotational levels is divided into four subsystems: (A, $\tilde{B}_1$), ($\tilde{A}$, B_1), (B_2, $\tilde{B}_3$), ($\tilde{B}_2$, B_3), between which collisional transitions are not allowed.

In D group classification, a distinction is not made between the levels with even and odd values of the rotational quantum number J. In classification according to the group D_{2h}, the states have the same mathematical symbols as in the previous classification, but the levels with odd values of J are denoted by a tilde (~).

The different description of the rotational states of water molecules during collisions with an atom and molecule of the linear rotator variety on the one hand, during interaction with radiation on the other, has a physical consequence. Radiative transitions turn out to be allowed between collisionally isolated subsystems $(A, \tilde{B}_1) \rightleftarrows (\tilde{A}, B_1)$, $(B_2, \tilde{B}_3) \rightleftarrows (\tilde{B}_2, B_3)$. This is the first example in molecular kinetics when the transitions during collisions do not repeat the radiative transitions and turn out to be more tightly constrained.

In experiments for verifying the collisional selection rules, heavy water vapor at room temperature was used in a mixture with argon (see Sect. 5 of Ref. [10], and also Ref. [11]). The concentration of the water was chosen to be small in comparison with the buffer gas concentration so that water-water collisions, for which the same rotational transitions are allowed as in the interaction with radiation, were rare in comparison with the frequency of water-argon collisions. The transition between the collisionally isolated subsystems $(A, \tilde{B}_1)$, $(\tilde{A}, B_1)$ was saturated by submillimeter radiation and the time to establish an equilibrium population (relaxation time) was measured as a function of buffer gas pressure. It turned out that increasing the frequency of the water-argon collisions by an order of magnitude did not have an effect on the relaxation time. This means that, in relation to the water molecule-argon atom interaction, there actually are collisionally isolated subsystems of rotational levels in molecular water.

There is an experiment that was performed in 1966 relating to double microwave resonance in the rotational spectrum of ethylene oxide molecules, which also indicates that this molecule has a collisionally isolated subsystem of rotational levels [12]. The effect of the prohibition on collisional transitions between subsystems is observed in pure ethylene oxide without a buffer gas, which is possible if the RL-interaction potential is significantly greater than the RR-potential. Ethylene oxide molecules (likewise an asymmetric rotator) have a dipole moment and a point group of symmetry C_{2v}.

In a theoretical paper [13], the interaction of water molecules with atoms for pairs of collisions in the gas phase is examined. The existence of a prohibition on the collisional molecule-atom transitions is predicted, which will lead to the emergence of a collisionally isolated subsystem of rotational levels in molecular water. The intrinsic rotation of the molecule is described using D-functions, the angular portion of the coordinate motion of the molecule-atom system is described using Y-functions, and a model interaction potential is used.

The derivation of the collisional selection rules is based on the concept of the conservation of spatial parity concurrently for both rotational motions. This approach raises objections, since, as was shown above, a D-function is not transformed by an irreducible group representation of spatial inversion and, in the case of the D-description, the concept of rotational state parity does not exist.

Reference [14] introduces the Y-description of the rotational motion of an asymmetric rotator and, using the spatial shape of the molecule, the collisional selection rules are predicted by the group-theory method for molecules with symmetry groups D_{2h}, C_{2v}, C_s. Examples of such molecules could be ethylene, water, and semiheavy water, respectively.

The author of a theoretical work [15], where the collisional selection rules during water molecules interacting with argon atoms are discussed along with the experimental data of Ref. [11], starts from the same initial premise that notions of the spatial shape of water molecules should be the foundation of the selection rules. However, the theoretical methods which are used in realizing this idea are questionable.

It is not possible to agree with the fundamental premise of the work in that "the point symmetry of the molecules imposes constraints on the form of the generalized multipole expansions of the interaction potential." As was shown in Sect 3, the symmetry group of the molecule does not act on a set of points of a three-dimensional sphere, where the generalized multipole potentials are defined, and there is no direct connection between the symmetry group of the molecule and the group of transformations in a set of angles (φ, ϑ, ψ), relative to which some multipole potential is invariant.

The second contradiction of the work consists in the fact when symmetry group operations act on generalized potentials, it is not the D-functions that are transformed, in terms of which the generalized potentials are written, but the several coefficients β in the expansion of the potentials, which play the role of additional variables. The rotational wave functions do not have these coefficients, which are derived from solving the eigen value problem with the Hamiltonian of a symmetric and asymmetric rotator. The application of group theory to establishing selection rules is based on the potential's matrix elements vanishing due to the symmetry of the wave functions and the potential function relative to the transformation of their common arguments, over which the integration is performed. Here the integration is done over angles φ, ϑ,

ψ, and the variable coefficients β are subject to transformation. Integration is not performed over these latter coefficients. They remain in writing the matrix element, and then additional analysis is done in order to establish which of the expressions vanish. This procedure is found to be in contradiction with the Wigner method of applying group theory in the problems of quantum mechanics [4].

4. Fixed and Rotatable Coordinate Systems in Quantum Rotator Theory

This section is devoted to clarifying the question of how the infinitesimal operators of the right and left regular group representations *SU*(2) and *SO*(3) [3, 16, 17] and the angular momentum operators in the fixed (laboratory) and rotating (molecular) coordinate systems [5, 18] are related to each other. This is because the physical literature assumes that the commutation relations of the operators in the rotational coordinate systems differ in sign from the commutation relations of the infinitesimal operators and the operators in the fixed coordinate system. The question of the sign of the commutation relations has a methodological nature, since the consequences of the physical nature of this difference are not yet known.

The infinitesimal operators represent the basis vectors of algebra *su*(2), for which the bracket multiplication of vectors follows from the structural constants of a Lie group, therefore the infinitesimal operators of any representation have the same commutation relations independent of their realization. The operators of orbital L, spin I, and total J momentum differ from the infinitesimal operators of the representations in spherical function space, spinor space, and D-function space, respectively, by a numerical multiplier $i\hbar$; they therefore have identical commutation relations, coinciding with those for infinitesimal operators.

Since D-functions, being the matrix elements of irreducible group representations *SU*(2), are given as functions in the group, then the infinitesimal operators of right A_k and left B_k, k = 1, 2, 3, regular representations turn out to be defined for them. The operators A_k and $B_{k'}$ mutually commute for any k and k'. In Ref. [3], a method is given for calculating the operators of the regular representation in the form of differential expressions of the Euler angles. Using Refs. [3, 8], we have:

$$A_1 = (\sin\vartheta)^{-1}\sin\psi\,\partial_\varphi + \cos\psi\,\partial_\vartheta - \operatorname{ctg}\vartheta\sin\psi\,\partial_\psi\,,$$
$$A_2 = (\sin\vartheta)^{-1}\cos\psi\,\partial_\varphi - \sin\psi\,\partial_\vartheta - \operatorname{ctg}\vartheta\cos\psi\,\partial_\psi\,, \quad A_3 = \partial_\psi\,, \tag{5}$$

$$B_1 = \operatorname{ctg}\vartheta \sin\varphi\partial_\varphi - \cos\varphi\partial_\vartheta - (\sin\vartheta)^{-1}\sin\varphi\partial_\psi,$$
$$B_2 = -\operatorname{ctg}\vartheta\cos\varphi\partial_\varphi - \sin\varphi\partial_\vartheta + (\sin\vartheta)^{-1}\cos\varphi\partial_\psi, \quad B_3 = -\partial_\varphi. \tag{6}$$

In the physical literature [5, 9, 17, 19], the following infinitesimal operators are used for the fixed coordinate system (we further assume that $h = 1$):

$$(-iJ_x) = \operatorname{ctg}\vartheta\cos\varphi\partial_\varphi + \sin\varphi\partial_\vartheta - (\sin\vartheta)^{-1}\cos\varphi\partial_\psi,$$
$$(-iJ_y) = \operatorname{ctg}\vartheta\sin\varphi\partial_\varphi - \cos\varphi\partial_\vartheta - (\sin\vartheta)^{-1}\sin\varphi\partial_\psi, \quad (-iJ_z) = -\partial_\varphi.$$

Complete matching of the operators $(-iJ_x)$, $(-iJ_y)$, $(-iJ_z)$ with the operator of the left representation B_k, $k = 1, 2, 3$, is achieved by the automorphism $u' = u_0 u u_0^{-1}$ of the $SU(2)$ group using the matrix

$$u_0 = \begin{Vmatrix} \alpha & 0 \\ 0 & \bar{\alpha} \end{Vmatrix}, \qquad \alpha = \exp(-i\pi/4).$$

The algebra of $su(2)$ and the algebra of the infinitesimal operators also undergo automorphic reflection, which is represented as a transformation of basis vectors of the algebra and its operator representations:

$$\| a'_1, a'_2, a'_3 \| = \| a_1, a_2, a_3 \| \begin{Vmatrix} 0 & 1 & 0 \\ -1 & 0 & 0 \\ 0 & 0 & 1 \end{Vmatrix}.$$

Here the vectors (a'_1, a'_2, a'_3) correspond to the operators $(-iJ_x)$, $(-iJ_y)$, $(-iJ_z)$, and the vectors (a_1, a_2, a_3) correspond to operators B_1, B_2, B_3.

A similar correspondence can be found between the infinitesimal operators of the moving coordinate system [18, 19]

$$(-iJ'_x) = (\sin\vartheta)^{-1}\cos\psi\,\partial_\varphi - \sin\psi\,\partial_\vartheta - \operatorname{ctg}\vartheta\cos\psi\,\partial_\psi,$$
$$(-iJ'_y) = -(\sin\vartheta)^{-1}\sin\psi\,\partial_\varphi - \cos\psi\,\partial_\vartheta + \operatorname{ctg}\vartheta\sin\psi\,\partial_\psi, \quad (-iJ'_z) = -\partial_\psi$$

and the infinitesimal operators of the first regular representation A_k:

$$(-iJ'_x) = A_2, \quad (-iJ'_y) = -A_1, \quad (-iJ'_z) = -A_3.$$

From this, it follows that the angular momentum operators of the moving coordinate system are expressed in terms of the A_k operators, but have an additional factor of (-1), naturally giving rise to the factor of (-1) in their commutation relations. Thus, the operators of the moving (primed) coordinate system, except for sign, are essentially no different from the infinitesimal operators of the right regular representation.

Works are known [19, 20] which discuss the procedure for deriving the angular momentum operators in the moving coordinate system starting from the operators in the fixed coordinate system. In the method for projecting the angular momentum operators onto the coordinate axes, rigidly fixed to the molecular axes [20], relations are used between the momentum operator J and the direction cosines λ, for example:

$$J_x = \lambda_{xX} J_X + \lambda_{xY} J_Y + \lambda_{xZ} J_Z, \tag{7}$$

$$\lambda_{xX} J_Y - J_Y \lambda_{xX} = i\lambda_{xZ}. \tag{8}$$

Here the indices X, Y, Z refer to the laboratory (fixed) coordinate system, and the indices x, y, z refer to the molecular (moving) coordinate system. We shall show that Eqs. (7) and (8) are mutually contradictory.

Transformation of the infinitesimal operators of the form of Eq. (7) arise only for automorphic reflection of group $SU(2)$ using some arbitrary but constant matrix $u_0 \in SU(2)$. The Euler angles $(\varphi_0, \vartheta_0, \psi_0)$ assign the transformation matrix $u_0 = u(\varphi_0, \vartheta_0, \psi_0)$; the direction cosines depend on these same angles. This transformation, for example, was used above when going from B_1 B_2, B_3 operators to $(-iJ_x)$, $(-iJ_y)$, $(-iJ_z)$ operators. Since $\lambda_{xX}(\varphi_0, \vartheta_0, \psi_0)$ is a constant quantity, then the second term on the left hand side of Eq. (8) must vanish, and the left and right hand sides of Eq. (8) turn out to be equal to each other.

We shall assume, as in Refs. [19, 20], that the direction cosines are functions of the angles $(\varphi, \vartheta, \psi)$, on which depend the trigonometric factors in the reflection operators A_k, B_k and with respect to which the differentiation is performed. In this case, all terms on the right hand side of Eq. (7) and the first term on the left hand side of Eq. (8) do not belong to the space of the infinitesimal operators, representing algebra $su(2)$, and their meaning cannot be established while remaining in the framework of $SU(2)$ group theory representations. This contradiction makes the projection method incorrect from the theoretical point of view. The differential operator obtained by this method cannot be brought into correspondence with the infinitesimal operators of any $SU(2)$ group representations.

In conclusion, we point out that the operators A_k, B_k possess the necessary properties for them to be used in molecular theory in the spirit of Ref. [20].

5. Conclusion

Two forms of describing the rotational motion of polyatomic molecules, which was addressed above, can be related to the fact that there is (Y-description) and is not (D-description) parity among the quantum numbers that determine the rotational state of the molecule. When interacting with a quantum-mechanical system, which certainly have parity (atom, diatomic molecule, linear rotator type molecule), a natural requirement consists in a polyatomic molecule also having a description with parity. A similar situation is encountered in the problem of electromagnetic wave radiation. If the system has parity (atom, atomic nucleus), then an expansion of the electromagnetic field in terms of multipoles and the wave function of a photon with a definite angular momentum and parity are used.

REFERENCES

1. L. S. Pontryagin, "Nepereryvnye gruppy [Continuous Groups]", Moscow, Nauka, 1984, 520 pp.

2. L. S. Pontryagin, "Obobshcheniya chisel [Unification of Numbers]", Moscow, Nauka, 1986, 117 pp.

3. N. Ya. Vilenkin, "Spetsial'nye funktsii i teoriya predstavlenii grupp [Special Functions and Group Theory Representation]", Moscow, Nauka, 1965, 588 pp.

4. E. P. Wigner, "Group Theory and its Application to the Quantum Mechanics of Atomic Spectra", New York, Academic Press, 1959.

5. L. D. Landau and E. M. Lifshits, "Quantum Mechanics: Nonrelativistic Theory, New York, Pergamon, 1977.

6. T. Oka, "The parity of rotational levels", J. MOL. SPECTROSC., Vol 48 pp 503-507, 1973.

7. A. S. Bruev, "Subbarrier collision transitions in the rotational spectrum of polyatomic molecules", KHIM. FIZ., Vol 5 No 2 pp 165-174, 1986.

8. V. K. Konyukhov, "Fixed and rotated coordinate systems in

quantum rotator theory", Preprint IOFAN No 19, Moscow, 1984, 33 pp.

9. D. A. Barshalovich, A. N. Moskalev, and V. K. Khersonskii, "Kvantovaya teoriya uglovogo momenta [Quantum Theory of Angular Momentum]", Leningrad, Nauka, 1975, 436 pp.

10. V. I. Tikhonov, "Rotationally nonequilibrium processes in water and heavy water vapor under supersonic flow conditions", this volume.

11. S. S. Bakastov, V. K. Konyukhov, and V. I. Tikhonov, "Kinetically isolated subsystems of rotational levels of heavy water molecules during atomic-molecular collisions", PIS'MA ZH. EKSP. TEOR. FIZ., Vol 37 No 9 pp 427-429, 1983.

12. T. Oka, "Collision-induced transitions between rotational levels", ADV. ATOM. MOL. PHYS., Vol 9 pp 127-206, 1973.

13. V. I. Selyakov, "Probability of rotational transitions of water molecules during collisions with atoms", ZH. PRIKL. MEKH. TEKH. FIZ., No 3 pp 10-20, 1980.

14. V. K. Konyukhov, "Collision selection rule in the rotational spectrum for asymmetric rotator type molecules", KRAT. SOOBSHCH. FIZ., No 10 pp 20-23, 1982.

15. A. S. Bruev, "Collision selection rule for asymmetric rotator type molecules", ZH. EKSP. TEOR. FIZ., Vol 86 No 6 pp 2056-2060, 1984.

16. D. P. Zhelobenko and A. I. Shtern, "Predstavleniya grupp Li [Lie Group Representation]", Moscow, Nauka, 1983, 360 pp.

17. A. N. Leznov and M. V. Savel'ev, "Gruppovye metody integrirovaniya nelineinykh dinamicheskikh sistem [Group Methods for Integrating Nonlinear Dynamic Systems]", Moscow, Nauka, 1985, 279 pp.

18. L. C. Biedenharn and J. D. Lauck, "Angular Momentum in Quantum Physics", Reading, Mass, Addison-Wesley, 1981.

19. A. G. Sitenko and V. K. Tartakovskii, "Lektsii po teorii yadra [Lectures on Nuclear Theory]", Moscow, Atomizdat, 1972, 285 pp.

20. Y. H. Van Vleck, "The coupling of angular momentum in molecules", REV. MOD. PHYS., Vol 23 pp 213-227, 1951.

SEMI-ANALYTICAL METHOD OF STUDYING THE KINETICS OF ROTATIONALLY NONEQUILIBRIUM FLOW IN JETS AND NOZZLES

A. V. Bogdanov, Yu. E. Gorbachev, and N. V. Stankus

0. Introduction

Analytical and semi-analytical methods for solving the kinetic equations for molecular populations [1-8] have great value both for modeling the basic features of molecular systems in jets, and for analyzing the operation of electronically ionized and gas-dynamic lasers. The majority of these methods is based on the diffusion approximation and is adapted to describing the kinetics of vibrational populations. The semi-analytical method is favorably different from its predecessors. This method was developed for the isothermal case in Refs. [9, 10] and generalized to the case of time-dependent macroscopic parameters in Ref. [11]. The main advantage of this approach - separation of the equations for the modified populations - makes it extremely attractive for use in calculating the kinetics of vibrational-rotational populations of complex molecular systems. The technique permits one to compute the evolution of only a few intermediate values of the population and to extrapolate the remaining values for sufficiently smooth distributions. This also saves computer memory and calculation time.

However, the direct application of the method in Refs. [9-11] to analyze the kinetics of rotational populations has a series of complications. First, the rotational excitation rate constant does not vary monotonically as a function of the initial quantum number, and therefore the "relaxation time" for some of the levels can be negative. Consequently, changing the variables, as suggested in Ref. [9], in this situation leads to an equation which becomes unstable when discarding the source term, and the required separation is not achieved. Second, in rotational kinetics due to the rotational quantum being so small, multi-quantum exchange turns out to be important; therefore, the approach taken in Refs. [9-11] and based on the assumption of single quantum transitions, needs to be modified. Third, there is a significant difference between vibrational and rotational spectra, which is related to the strong nonequivalency and degeneracy of the latter. In connection with this, the problem of going from "adiabatic" variables to original populations becomes significantly more complex in the theory of rotational relaxation.

As we shall see below, all of the aforementioned difficulties are completely compensated by the simplicity and clear representation of describing the kinetics in "adiabatic" variables. In particular, we shall show that the entire region of the jet is naturally divided into three portions, in each of which simple expressions can be derived for solving in integral terms from some combination of rate constants along the flow tube. As a result, in the proposed approach it is very simple to overcome the familiar difficulties related to describing the deviation of the solution from the equilibrium state. Most interesting is the region where the modified populations have quasi-stationary portion (plateau). The position of this plateau is related to the rotational excitation rate constants in a system of purely algebraic relations. This circumstance reveals additional possibilities for solving the reverse problem of relaxation spectroscopy.

1. Conversion to "Adiabatic" Variables

We shall demonstrate the different means of going from individual populations of rotational states $c_j(t)$ ($\sum_j c_j(t) = 1$) to adiabatic variables on a simple example of RT-relaxation in the single-quantum approximation. The equations for $c_j(t)$ have the form

$$\dot{c}_j(t) = I_{j-1}(t) - I_j(t), \tag{1}$$

and for the quantum flux I_j in the simple situation being examined, we have the expression

$$I_j(t) = K_{j,j+1} c_j(t) - K_{j+1,j} c_{j+1}(t), \tag{2}$$

where $K_{i,j}$ are the rates constants of the transition from state i to state j, depending only on the translational temperature and quantum numbers. Introducing the ratio of upward and downward transition rate constants from a given level

$$a_{j-1}(t) = K_{j-1,j}/K_{j,j-1}, \tag{3}$$

it is convenient to define the modified population $f_i(t)$ as the ratio [9]

$$c_j(t)/c_{j-1}(t) = a_{j-1}(t) f_{j-1}(t). \tag{4}$$

The contribution to $\dot{f}_j$ consists of three parts [11]:

$$\dot{f}_j = R(f_j, f_j) + H_j f_j + S_j(f_k) f_j. \tag{5}$$

Here $R(f_j, f_j)$ is the quadratic relaxation term:

$$R(f_j, f_j) = A_j f_j^2 + B_j f_j + C_j, \tag{6}$$

which vanishes at the equilibrium solution ($f_j^0 = 1$) by virtue of the relation

$$A_j + B_j + C_j \equiv 0, \quad A_j = K_{j+1,j+2} - K_{j,j+1}, \quad C_j = K_{j+1,j} - K_{j,j-1}; \tag{7}$$

$H_j(t)$ determines the contribution of the gas-dynamic gradients to the relaxation:

$$H_j(t) = -d \ \ln a_j(t)/dt, \tag{8}$$

and $S_j(f_k)$ is related to the presence of pumping "sources" in the flow:

$$S_j(f_k) = K_{j,j-1}(1 - f_j f_{j-1}^{-1}) f_j^{-1} - K_{j+1,j+2}(1 - f_{j+1} f_j^{-1}) f_j. \tag{9}$$

In the absence of pumping, this is a term that is second order in relation to both the deviation from equilibrium and the smoothness of the initial distribution. In general then in the jet

$$A, B, C \gg H \gg S, \tag{10}$$

therefore, we did not consider S in our studies. Note, however, that the difference equations (5) are coupled only through $S(f_k)$, from which it follows that transformation (4) plays a role similar to converting from phase to adiabatic variables in Hamiltonian mechanics [12]. The equations for $f_j(t)$, written in the form of Eq. (5), are much more useful than Eqs. (1) in practical calculations.

For further investigation, the function

$$\tau_j = (C_j - A_j)^{-1}, \tag{11}$$

has very important significance, which represents the relaxation time of a given level j [11].

Analysis of the stability of Eq. (5) in linear approximation shows that its solution is stable for small perturbations and, consequently, the $S_j(f_k)$ terms can be justifiably disregarded when the condition $\tau_j > 0$ is satisfied. If $\tau_j < 0$, changing the variables in Eq. (4) does not lead to simplifying the problem. If for level i the quantity $\tau_i < 0$, then, as it turns out, the role of the "adiabatic" variables is played by the functions

$$g_i = \frac{I_i}{I_{i-1}} \equiv \frac{z_{i+1} - z_i}{z_i - z_{i-1}}\, b_i, \tag{12}$$

where b_i is the combination of rate constants:

$$b_i = a_i K_{i+1,i}/K_{i,i-1}, \tag{13}$$

and $z_i = c_i/c_i^e$, where c_i^e is determined from the equation $I_i = 0$. Since $f_{j-1} = z_j/z_{j-1}$, then, combining Eqs. (5) for $j = i$ and $j = i + 1$, we obtain an equation for g_i [10]:

$$\dot{g}_i = \tilde{R}(g_i, g_i) + \tilde{H}_i(g_i) + \tilde{S}_i(g_k)\, g_i, \tag{14}$$

i.e., an equation of the same type as Eq. (5), where

$$\begin{aligned} &\tilde{R}(g_i, g_i) = \tilde{A}_i g_i^2 + \tilde{B}_i g_i + \tilde{C}_i, \\ &\tilde{A}_i = K_{i+1,i} - K_{i,i-1}, \quad \tilde{C}_i = K_{i,i+1} - K_{i-1,i}, \\ &\tilde{S}_i = K_{i,i-1}(g_i g_{i-1}^{-1} - 1)\, g_i^{-1} - K_{i+1,i}(1 - g_{i+1} g_i^{-1})\, g_i. \end{aligned} \tag{15}$$

It is not difficult to see that the signs of τ_j and $\tilde{\tau}_j = (\tilde{C}_j - \tilde{A}_j)^{-1}$ are opposite. Therefore, for those equations for which the f_j terms become unstable, the solution for g_i has that property.

The formula for $\tilde{H}_i$ has the form

$$\tilde{H}_i = \dot{b}_i/b_i - (f_i - 1)^{-1} f_{i-1}^{-1}\, [(\dot{a}_i/a_i) f_i f_{i-1} - (\dot{a}_{i-1}/a_{i-1})\, g_i/b_i]. \tag{16}$$

By virtue of Eq. (13), b_i differs from a_i by a factor that is weakly dependent on temperature [13], which can be used to simplify Eq. (16).

We shall examine the question of the reverse conversion, i.e., from "adiabatic" variables to the original variables. We suggest that at all levels up to level j^*, the value of $\tau_j > 0$. Then by virtue of Eq. (3)

$$c_j(t) = c_0 \prod_{i=0}^{j-1} a_i f_i(t), \quad j \leqslant j^*.$$

The principle of detailed equilibrium

$$a_j = K_{j,j+1}/K_{j+1,j} = [s(j+1)/s(j)] \exp\{-[E(j+1)-E(j)]/kT\},$$

where $E(j)$ is the energy of state j, and $s(j)$ is its statistical weight, gives

$$\prod_{i=0}^{j-1} a_i = s(j) \exp[-E(j)/kT].$$

We finally obtain the conversion to the old variables in the form

$$c_j(t) = c_0 s(j) \exp[-E(j)/kT] \prod_{i=0}^{j-1} f_i(t). \tag{17}$$

Since the molecules in the rotational ground state, as a rule, are in equilibrium with the translational degrees of freedom, then it can be assumed that $c_0 = Q^{-1}$, where Q is the statistical sum which eliminates the question of finding c_0. The expression for Q is obtained if the normalization condition for c_i^e is rewritten in the form

$$Q = \sum_j s(j) \exp[-E(j)/kT].$$

If for $j > j^*$ the value of $\tau_j < 0$, then the subsequent calculations are done in terms of g_i, and the recursion formula has the form

$$z_j = z_{j^*-1} + (z_{j^*} - z_{j^*-1}) \sum_{i=j^*}^{j-1} \prod_{k=j^*}^{i} g_k/b_k. \tag{18}$$

If with an increase in j the value of τ_j again goes positive, then in this region it is necessary to convert to f-variables.

2. Features of Modeling RR-Exchange and Multi-Quantum RT-Exchange

It is customary to assume that, due to the rotational spectrum being strongly non-equidistant, the role of resonance effects in *RR*-exchange is small and the *RR*-exchange constants do not make too great a contribution in modeling the jet. Analysis [14], however, shows that precise quantum-mechanical calculations [15] confirm that *RR*-exchange also occurs efficiently when there is partial resonance in the colliding partners, and such processes can also make a noticeable contribution to the kinetics. A series of works [16-18] analyzed the contribution of *RR*-exchange and multi-quantum exchange to the evolution of the populations of the rotational degrees of freedom, however, to date there is no complete answer to the question of the how large these processes' contribution are and what sign they have. We shall show that, at least when the deviation from equilibrium is not too large in the kind of approach being examined here, a fairly general conclusion can be drawn for the rate constants.

In the presence of *RR*-exchange in Eqs. (1) and (2), we should make the substitution

$$K_{i,j} \to K^{*}_{i,j} = K_{i,j} + n \sum_{k} K^{k,k+1}_{i,j} c_k. \tag{19}$$

Introducing "adiabatic" variables can be done by the same scheme, but with the one difference that the coefficients in Eqs. (5) and (14) will now depend on the individual populations $c_k(t)$. In the usual situation, when the deviation from equilibrium is not too great (see Sect. 6), we can perform a linearization near the solution corresponding to the absence of *RR*-processes. Then, in the expression for the inverse relaxation time (Eq. (11)), the corresponding adiabatic correction (Eq. (19)) is made.

Further simplifications are possible in a series of limiting situations. For $kT \ll B_e\bar{j}$, where $\bar{j}$ is the average effective value of j, the function inside the summation sign has a sharp maximum and Laplace's discrete method can be used to calculate the sum. In this case, in Eq. (19) instead of c_k enter $c_{\bar{j}}$, which can be calculated without much error for an equilibrium distribution. In the opposite limiting case ($kT \gg B_e\bar{j}$), we can replace the sum with an integral and express Eq. (19) through the moments of the distribution function, for which simple calculations can be used in terms of macroscopic parameters [2].

In any approximation, a corresponding correction to the inverse relaxation time of the level (Eq. (11)) will be an effective consequence of the presence of *RR*-exchange. This correction, when the deviation from equilibrium is not too great, when the treatment in Sect. 6 is applicable, leads to an acceleration of relaxation and, consequently, to a decrease in the maximum deviation from equilibrium.

Multi-quantum transitions are very important in the kinetics of rotational populations. This is because of both the small magnitude of the typical values of the rotational constant, and the features of the dependence of the rotational constants on the preassigned quantum number, i.e., the presence of nonmonotonic behavior with a maximum under certain conditions [13]. Calculation of multi-quantum exchange in Eqs. (1) leads to the following substitution in the formulas for the fluxes Eq. (2): $K_{j,j+1}c_j(t) \rightarrow \sum_\delta K_{j,j+\delta}c_j(t)$ (and similar for $K_{j+1,j}$). The general form of the substitution (4) in this situation has the form

$$c_{j+\delta}(t)/c_j(t) = \sum_{i=0}^{\delta-1} a_{j+i} f_{j+i}(t), \tag{20}$$

and the entire subsequent conclusion is completely analogous to that examined in Sect. 1, and again leads to an equation of the form of Eq. (5). However, the expression for the relaxation term is now different:

$$R(f_j, f_j) = \sum_\delta \left[A_{j,\delta} f_j^{\delta+1}(t) + B_{j,\delta} f_j(t) + C_{j,\delta} f_j^{1-\delta}(t)\right], \tag{21}$$

where (refining Ref. [11]) we have

$$A_{j,\delta} + B_{j,\delta} + C_{j,\delta} = 0. \tag{22}$$

The coefficients themselves have the simple form:

$$A_{j,\delta} = K_{j+1,j+\delta+1} - K_{j,j+\delta}, \quad C_{j,\delta} = K_{j+1,j+1-\delta} - K_{j,j-\delta}. \tag{23}$$

The terms that are second order in size relative to the smoothness of the distribution function and the deviation from the equilibrium position are collected in the "source" term.

Studying Eq. (5) for instability in this situation shows that the role of the relaxation time of level j is played by the quantity

$$\tau_j^{-1}(t) = \sum_\delta \delta\,(C_{j,\delta} - A_{j,\delta}), \tag{24}$$

converting to the old definition for $\delta = 1$. It turns out that in the diffusion approximation, this quantity is directly related to the diffusion coefficient with respect to levels. Thus, including multi-quantum exchange also leads to a decrease in the relaxation time, i.e, to a decrease in the deviation from equilibrium.

3. Flow in the Initial Layer. Perturbation Theory with Respect to the Relaxation Term

We shall begin the analysis of the kinetics of rotational relaxation with an initial layer in which the departure from equilibrium occurs. Although, as a rule, condition (10) is satisfied every where in the jet, the dominant term in the initial layer is $H_j f_j$. This is related to the fact that, first, the source term H_j is a maximum in the initial stage of the jet, and second, when departing equilibrium $f_j \simeq 1$ and in view of Eq. (22) $R(f_j, f_j) \simeq 0$. Therefore, in the initial stage we can solve Eq. (5) assuming the second term to be dominant and treating the first term by perturbation theory. Since

$$H_j = -\frac{d \ln a_j(t)}{dt} = \frac{d}{dt}\left(\frac{\Delta E_j}{kT}\right) = -\frac{\Delta E_j}{k}\frac{dT/dt}{T^2} \equiv -G(t)\Delta E_j, \tag{25}$$

in the zeroth order approximation we have the equation

$$\dot{f}_j(t) = -G(t)\Delta E_j f_j(t), \tag{26}$$

which with the equilibrium initial condition is easily integrated and gives

$$f_j^{(0)}(t) = \exp[-\Delta E_j \int_0^t G(t)dt]. \tag{27}$$

Since Eq. (27) is Green's function for the corresponding homogeneous equation, the corrections to Eq. (27) are derived on the basis of the formal representation

$$f_j^{(1)}(t) = \exp[-\Delta E_j \int_0^t G(t)dt] \int_0^t R(f_j, f_j)_\tau \exp[\Delta E_j \int_0^\tau G(t_1)dt_1] d\tau. \tag{28}$$

All of the integrals in Eqs. (27) and (28) are evaluated along the flow tube and are determined by solving the gas-dynamic equations. For example, the first approximation is derived from Eq. (28) by the substitution $R(f_j, f_j) \rightarrow -(f_j^{(0)} - 1)\tau_j^{-1}$. The structure of the following terms of the expansion are easily evident and derived form the expression in Sect. 5. Note that in the jet, the contribution $f_j^{(1)}$ is negative, and therefore Eq. (27) defines an upper limit for the deviation from equilibrium in a free jet. Allowing for all possible corrections decreases this deviation.

We shall also show that from Eq. (27) there is an obvious difference in the flow of a vibrationally nonequilibrium gas, for which ΔE is almost constant and slowly decays with j, and a rotationally nonequilibrium gas, when ΔE increases linearly with increasing quantum number. This situation does indeed correspond to strong nonequilibrium at high rotational levels.

In order to illustrate the behavior of the solution and simultaneously estimate the maximum errors introduced in our approximation approach, we shall examine the case of very large deviations from equilibrium, when the validity of disregarding the "source" terms is determined only by the smoothness of the initial distribution. Such deviations are realized when examining the kinetics of small impurities of polar molecules in an inert gas at moderate temperatures.

We shall examine the He-HF system, for which the deviations are maximum at fore-chamber temperatures of 300-500 K (at lower temperatures, the upper levels are rapidly "quenched"). The gas-dynamic flow was modeled in the approximation of a spherical point source [19], and analytical parameterization of the calculations by the method of classical trajectories [13] was used for the rate constants.

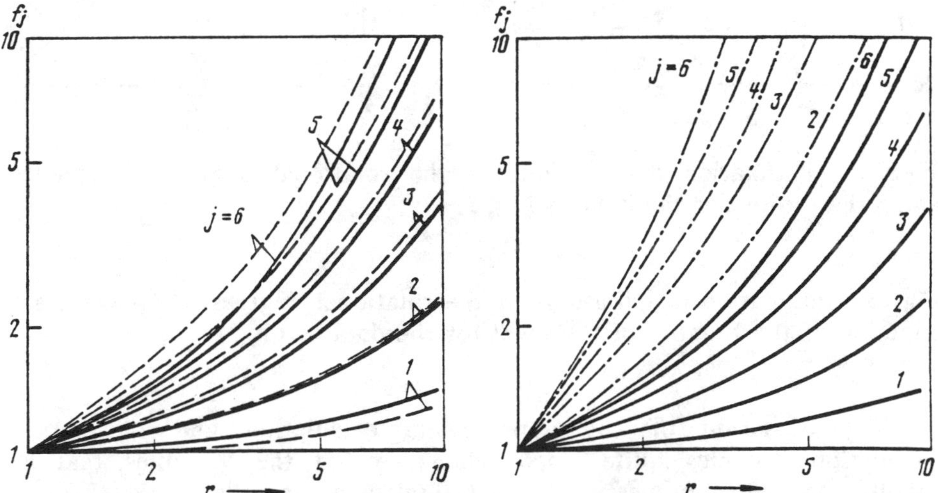

Fig. 1. Comparison of the solutions to the equations of the level kinetics (solid lines), Eqs. (5) without the source term (dashed line), and estimates of the maximum deviation from equilibrium based on Eq. (27) (dot-dashed line) for different j on a sample of the He-HF system with T_0 = 300 K.

Figure 1 shows the dependencies of f_j on the distance along the axis of the nozzle for a series of levels, calculated by exactly solving the kinetic equations (1) by solving Eq. (5) without the source term and estimating the maximum deviation by Eq. (27). We see that all the qualitative features, found as a result of the asymptotic analysis, are also confirmed in the case of large deviations. On the other hand, it is clear that for large deviations from equilibrium, the simple formulas derived (suitable for $|f_j - 1| \lesssim 1$) give only qualitative agreement with the precise calculations. However, if the resultant distribution is recalculated by Eq. (17) at the population $n_j = nc_j$, the agreement improves significantly (Fig. 2).

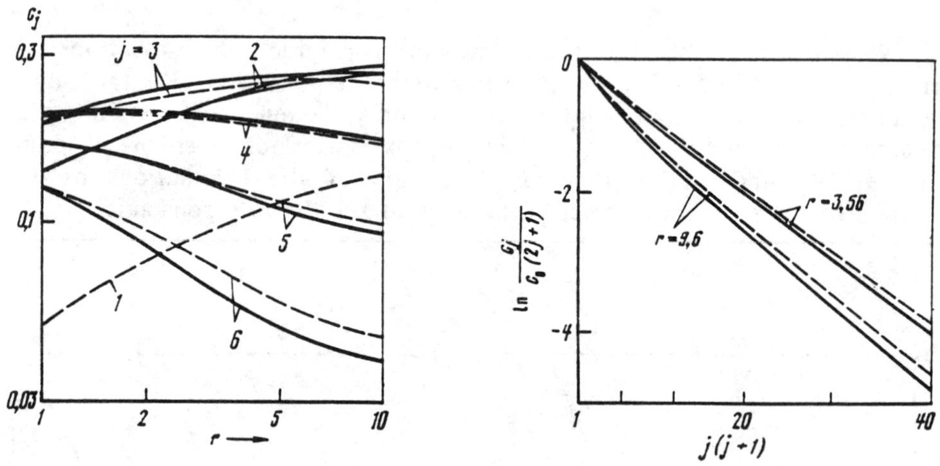

Fig. 2. Population distribution for the rotational degrees of freedom along the radius for the data in Fig. 1.

Fig. 3. Population distribution for the rotational degrees of freedom as a function of the number of the level for the data in Fig. 1.

The deviation of the approximate calculation and the precise calculation becomes quite small when we use the variables that are usually used when processing experimental data, namely, $\ln[c_j/c_0(2j + 1)]$ and $j(j + 1)$ (Fig. 3).

Thus, the following conclusion can be drawn from the comparison just performed: 1) in the variables in Eq. (4), the evolution of the populations in the initial stage has a simple form that is independent of the dynamics of the system and is determined only by its spectrum; 2) a simple analytical description can be derived for the evolution in the initial stage (quantitative for $|f_j - 1| \lesssim 1$ and qualitative for $|f_j| > 2$, however the range of applicability of this description in traditional variables is broadened substantially; 3) the traditional variables are slightly sensitive to the errors in calculating the kinetics, and therefore the finer questions for comparison, for example in the problem of relaxation spectroscopy (see Sect. 6), should be solved in the new variables.

4. Solution in the Relaxation Zone. Perturbation Theory with Respect to the Gas-Dynamic Gradients

The contributions of $R(f_j, f_j)$ and $H_j f_j$ to the kinetic equation (5) are compared relative to attaining the maximum deviation from equilibrium, i.e., for relaxation times on the order of τ_j. In what follows, however, H_j when getting further from the nozzle's slit continuous to decay rapidly and $R(f_j, f_j)$ becomes the leading term in the right hand side of the equation. The solution in the zeroth approximation is obtained by discarding H_j and reduces to the polynomial in Ref. [11]:

$$f_j^{(0)}(t) = \frac{1 + f_j' + (1 - f_j')\,[\int_{\tau_j}^{t} E(j, t')\Lambda(j, t')dt' - E(j, t)]}{1 + f_j' + (1 - f_j')\,[\int_{\tau_j}^{t} E(j, t')\Lambda(j, t')dt' + E(j, t)]}\,, \tag{29}$$

where f'_j is the value of $f_j(t)$ at $t = \tau_j$; $\Lambda(j) = A_j + C_j$, and $E(j, t)$ is the integral along the flow tube:

$$E(j, t) = \exp\,[\int_0^t dt'/\tau_j(t')\,]. \tag{30}$$

The following terms of the expansion with respect to the gas-dynamic gradients are found from the recurrence relations

$$\begin{aligned} \dot{f}_j^{(l)}(t) &= [2A_j f_j^{(0)}(t) + B_j]\, f_j^{(l)}(t) + \\ &+ A_j \sum_{k=1}^{l-1} f_j^{(k)}(t) f_j^{(l-k)}(t) + H_j(t) f_j^{(l-1)}(j, t). \end{aligned} \tag{31}$$

The solution is expressed in terms of the integral along the flow tube

$$D_j(t) \equiv \exp\left\{\int_{\tau_j}^{t} [2A_j f_j^{(0)}(t') + B_j]\,dt'\right\} \tag{32}$$

and has the form

$$f_j^{(l)}(t) = D_j(t) \int_{\tau_j}^{t} D_j^{-1}(t')[A_j \sum_{k=1}^{l-1} f_j^{(k)}(t') f_j^{(l-k)}(t') + H_j f_j^{(l-1)}(t')]dt'. \quad (33)$$

It is not difficult to see that the corrections to Eq. (33) are positive, and therefore the effect of the gas-dynamic gradients protract the relaxation. Since, however, for large $t >> \tau_j$, the function $f_j^{(0)} \simeq 0$, making allowance for the correction in Eq. (33) does not qualitatively alter the behavior of the solution for $t > \tau_j$. Thus, the solution for $t > \tau_j$ can evolve by one of two means. If the maximum deviation from equilibrium f'_j was not very large, i.e., $|f'_j - 1| \lesssim 1$, then for times on the order of $(2\text{-}3)\tau_j$ (note that τ_j also depends on t through the temperature dependence) $f_j(t) \to 1$ and the population of a given level becomes the equilibrium population, managing to be fine-tuned under the change of the temperature of the carrier gas. In the opposite case of large deviations ($f'_j >> 1$), the value of τ_j^{-1} approaches zero faster than f_j approaches unity, and quenching of the population occurs at temperatures corresponding to this t.

5. Small Nonequilibria. Linearized Solution

In order to illustrate all of the aforementioned features of the rotational relaxation processes and to derive simplified formulas for the important cases in practice of small deviations from equilibrium, we shall examine a linearized variant of Eq. (5), in which the relaxation term is replaced by the expression

$$R(f_j, f_j) = -\varphi_j(t)/\tau_j(t), \quad \varphi_j(t) = f_j(t) - 1. \quad (34)$$

The equation for $\varphi_j(t)$ has the form

$$\dot{\varphi}_j + L_j \varphi_j = H_j + Q_j(\varphi_j), \quad (35)$$

where $L_j = \tau_j^{-1} - H_j$ is a linear multiplication operator, and all quadratic correction terms which can be considered by the iterations are included in Q_j.

Since the initial conditions for φ_j are zero, the formal solution to Eq. (35) has the form

$$\varphi_j(t) = \exp[-\int_0^t L_j(t')dt'] \int_0^t [H_j + Q_j(\varphi_j)](\tau) \exp[\int_0^\tau L_j(T)dT] d\tau. \tag{36}$$

Equation (36) can serve as the basis for deriving the corrections in relation to the deviation from equilibrium for the iterations of the quadratic terms $Q_j(\varphi_j)$. In the linearized approximation, it is reasonable that $Q_j = 0$.

By way of expanding the exponential in Eq. (35), it is also possible to derive a perturbation theory series in the relaxation term, which is discussed in Sect. 3. Since for small deviations from equilibrium $H_j\tau_j < 1$, for simple calculations it is possible to disregard the term in L with the gas-dynamic gradients and to use the simplified relation

$$\varphi_j(t) = \exp[-\nu(t)] \int_0^t H_j(\tau) \exp[\nu(\tau)] d\tau, \tag{37}$$

$$\nu(t) = \int_0^t dt'/\tau_j(t'). \tag{38}$$

For a monotonically decaying function $H_j(t)$ as, for example, in a freely expanding jet, the structure of Eq. (37) can easily be understood even without performing calculations. At the initial stage from zero to $t = \tau_j$, the function φ_j grows from zero to a maximum value on the order of $H_j\tau_j$, and then monotonically decays to zero. We saw that, qualitatively, this pattern of relaxation takes place in the case of large deviations from equilibrium.

6. Quasi-Stationary Regime. Relaxation Spectroscopy

From the preceding analysis it is clear that in jet flows there is a very interesting region (where $t \sim \tau_j$) of slowly varying modified populations f_j (plateau). It arises due to the compensation of contributions to the right hand side of Eq. (5) for relaxation and gas-dynamic terms and is determined by the condition

$$A_j f_j^2 + B_j f_j + C_j + H_j f_j \simeq 0. \tag{39}$$

In accounting for multi-quantum exchange (Sect. 2), we get an algebraic equation of higher order.

The solution to Eq. (30) gives us the exact solution for the maximum deviation from resonance:

$$f_j^* = -(B_j + H_j)/2A_j + \sqrt{(B_j + H_j)^2/4A_j^2 - C_j/A_j} \tag{40}$$

(a minus sign in front of the radical corresponds to a nonphysical solution and should be omitted). This important relation relates the position of the maximum f_j and its height, i.e., the kinetic characteristics with the gas-dynamic flow. If H_j/A_j is assumed to be small, from Eq. (40) follows the approximate formula

$$\varphi_j^* = f_j^* - 1 = H(\tau_j)\tau_j, \tag{41}$$

in agreement with the results obtained from analysis of the simplified equation (38).

It should be emphasized that the location of the plateau is slightly sensitive to the form of the rate constants, and its value depends on them rather strongly. In addition, the position of the maximum will be the same for each level. If the parameters in the fore-chamber are chosen such that the maximum for neighboring levels corresponds to one τ (i.e., $\tau_j \approx \tau_{j+1}$), from Eq. (41) by virtue of the principle of detailed equilibrium and the definition in Eq. (6), follows the relation

$$K_{j+1,j}[1 - a_j - (1 - a_{j+1})^{-1}] - K_{j,j-1} = \tau_j^{-1} + \tau_{j+1}^{-1}(1 - a_{j-1})^{-1}, \tag{42}$$

relating the location of the maximum deviation from equilibrium, i.e. τ_j, with the values of the rate constants in the resulting system.

On the basis of the qualitative picture of the flow formulated here, the following procedure can be proposed to recover the rate constants of all transitions at some fixed temperature. By way of preliminary calculation of the flow with some model constants, pressures in the fore-chamber are established at which for the studied levels the maximum deviation from equilibrium (Eq. (41)) falls on that potion of the jet where a diagnostic aperture could be located. Due to the small sensitivity of the position of the maximum, the calculations in this stage can be estimates. In order for the maximum at all levels to correspond to a single temperature, the temperature in the fore-chamber during all measurements should be constant. A series of measurements (for determining the pressure for each level) gives us the values of the populations corresponding to the maximum deviation from equilibrium c_j^*. By virtue of Eq. (4), from here we find φ_j^* and, knowing $H_j(\tau_j)$ from gas-dynamic calculations, we obtain the value of τ_j from Eq. (24).

From system of Eqs. (42), it is now possible to sequentially determine all rate constants starting with $K_{1,0}$ ($K_{1,-1} \equiv 0$). The temperature at which the value of $K_{j+1,j}$ is determined depends on the position of the diagnostic aperture on the axis of the jet and is also found in the stage of the gas-dynamic calculation of the jet. Although a purposeful experiment has not yet been done in this formulation, the closest to it in the nature of the resulting data are experiments performed in The Gas-Dynamic Laser Laboratory [20]. There is reason to hope that in the near term simple scaling of these data will permit one to recover the rate constants of rotational excitation in the Ar-H_2O system.

7. Conclusion

Thus, the method examined for semi-analytical solutions to the population kinetics equations turns out to be rather fruitful both for the obvious analysis of the physical picture of kinetics, and also for quick and simple estimates of the evolution of the populations in free jets. We suggest that their value grows even more in the kinetics of vibrational-rotational relaxation due to the possibility of calculating only a few reference equations and interpolating the resulting data on the remaining values, and also by virtue of the high efficiency of the scheme of isolating the vibrational and rotational kinetics based on "adiabatic" variables. Additional advantages of the proposed method are related to the existence of an analytical formula that describes the departure from a state of equilibrium and permits one to save significant calculation time on a computer.

REFERENCES

1. B. F. Gordiets, A. I. Osipov, and L. A. Shelepin, "Kineticheskie protsessy v gazakh i molekulyarnye lazery [Kinetic Processes in Gases and Molecular Lasers]", Moscow, Nauka, 1980, 512 pp.

2. E. E. Nikitin and A. I. Osipov, "Kintetika i kataliz [Kinetics and Catalysis]", "T 4 Kolebatelnaya relaksatsiya v gazakh [Volume 4: Vibrational Relaxation in Gases]", Moscow, VINITI, 1977, 283 pp.

3. S. A. Resetnyak and L. A. Shelepin, "Diffusion Theory of Chemical Reactions", ZH. PRIKL. MEKH. TEKH. FIZ., No 5 pp 83-92, 1981.

4. S. A. Zhdanok, R. I. Soloukhin, and S. M. Khizhnyak, "Analytical theory of a CO-GDL and its application to analysis of the gas-dynamic

means of producing inversion in carbon oxides", Preprint ITMO No 7, Minsk, 1985, 26 pp.

5. S. A. Brau, "Classical theory of vibrational relaxation of anharmonic oscillators", PHYSICA, Vol 58 No 4 pp 533-553, 1972.

6. M. B. Zheleznyak and G. V. Naidis, "Vibrational level distribution of vibrational relaxation and dissociation rates of diatomic molecules under nonequilibrium conditions", ZH. PRIKL. MEKH. TEKH. FIZ., No 1 pp 4-9, 1976.

7. T. V. Bystrova, "Continuous description of vibrational relaxation of multilevel diatomic molecules", Preprint IPM AN SSSR No 238, Moscow, 1984, 52 pp.

8. G. I. Sukhinin, "Relaxation representation of kinetic equations", Preprint ITF SO AN SSSR No 114, Novosibirsk, 1986, 26 pp.

9. A. P. Vasil'ev, G. V. Dubrovskii, and V. M. Strel'chenya, "Approximate analytical description of vibrational relaxation of weakly anharmonic oscillators", ZH. PRIKL. MEKH. TEKH. FIZ., No 5 pp 16-24, 1984.

10. A. V. Bogdanov, Yu. E. Gorbachev, and V. M. Strel'chenya, "Relaxation and condensation processes influence on flow dynamics", in: Proc. XV Intern. Symp. on Rarefied Gas Dynamics", Grado (Italy), 1986, pp 407-408.

11. G. V. Dubrovskii and V. M. Strel'chenya, "Relaxation of anharmonic molecules", ZH. PRIKL. MEKH. TEKH. FIZ., No 3 pp 22-31, 1986.

12. C. V. Heer, "Statistical Mechanics, Kinetic Theory, and Stochastic Processes", New York, Academic Press, 1976.

13. A. V. Bogdanov, G. V. Dubrovskii, and A. I. Osipov, "Rotational excitation of diatomic and polyatomic molecules during collisions", KHIM. FIZ., Vol 4 No 9 pp 1155-1173, 1985.

14. A. V. Bogdanov and Yu. E. Gorbachev, "Nature of resonances in exchange processes of vibrational-rotational-translation energy", PIS'MA ZH. TEKH. FIZ., Vol 10 No 4 pp 234-237, 1984.

15. M. H. Alexander, "Close-coupling studies of rotationally inelastic HF-HF collisions at hyperthermal energies", J. CHEM. PHYS. Vol 75 No 10

pp 5135-5146, 1980.

16. Koura "Rotational distribution of N_2 in Ar free jet", PHYS. FLUIDS, Vol 24 No 3 pp 401-405, 1981.

17. P. A. Skovorodoko and R. G. Sharafutdinov, "Kinetics of the populations of rotational levels in a free nitrogen jet", ZH. PRIKL. MEKH. TEKH. FIZ., No 5 pp 40-49, 1981.

18. G. V. Dubrovskii, A. V. Bogdanov, and R. E'. Mukhametdzyanov, "Rotational relaxation of N_2 in free jets", ZH. TEKH. FIZ., Vol 54 No 6 pp 1196-1198, 1984.

19. H. Ashkenas and S. S. Sherman, "The structure and utilization of supersonic free jets in low-density wind tunnel", in: "Proc. IV Symp. Rarefied Gas Dynamics", New York, Academic Press, 1966, Vol 2, pp 84-105.

20. F. S. Bakastrov, V. K. Konyukhov, and V. I. Tikhonov, "Nonequilibrium rotational distribution function for D_2O molecules in a rarefied jet", ZH. PRIKL. MEKH. TEKH. FIZ., No 3 pp 28-35, 1985.

ORTHO-PARA-SELECTIVITY IN THE ADSORPTION OF H_2O AND D_2O MOLECULES FROM THE GAS-PHASE ONTO A SOLID SURFACE

V. K. Konyukhov, V. I. Tikhonov, and T. L. Tikhonova

The transition a molecule undergoes from a state of free motion in the gas phase to a bound state on the surface of a solid (which occurs in the process of adsorption and condensation of gases) is often accompanied by the violation of statistical equilibrium between the spin-modifications of the molecules. One modification goes predominantly into the bound state, and the other remains in the gas phase. Examples of this kind of process are: *ortho-para*-selective adsorption of hydrogen on an aluminum oxide surface at a temperature of 20 K [1, 2], *ortho-para*-selective condensation of D_2O molecules in a supersonic jet of carbon dioxide gas [3], the nonequilibrium population relative to spin-modifications of NH_3 molecules under conditions of a condensing supersonic jet [4], and spin-modification nonequilibrium H_2-HF dimers during their formation likewise under supersonic flow conditions [5].

The aim of this work was to experimentally verify whether there is spin-modification selectivity during the adsorption of water molecules on an aluminum oxide surface at room temperature. The adsorption of water on a surface is a widespread phenomenon in nature and in technological processes, such that the observation of *ortho-para*-selectivity of water adsorption is of general interest for physicochemical and biological processes.

Another part of this work is devoted to clarifying whether there is *ortho-para*-selectivity in the adsorption of D_2O molecules on a surface of solid carbon-dioxide. This study is undertaken in relation to the rotationally selective condensation effect. Earlier [3], rotational selectivity was explained as a spatial effect, arising in the initial stage of condensation with the formation of D_2O-CO_2 dimers. Another possible mechanism of rotational selectivity is experimentally demonstrated here, which is related to the adsorption of heavy water molecules on the surface of carbon dioxide gas clusters.

A short summary of this work was published in Refs. [6, 7].

Experiments on the adsorption of water vapor were done in a system of two cells connected by a short nozzle. One cell was made of Duralumin with a volume of 2.4 litres filled with corundum beads (α-modification Al_2O_3) 2 cm in diameter. The other cell was made of stainless steel with an internal diameter of 5 cm and 60 cm long and had teflon windows at the end with a total area of 40 cm^2 transparent in the submillimeter spectral region. The bulk concentration of water vapor was determined by the magnitude of the total absorption coefficient on the rotational transitions 4_{13}-5_{24} (*para*-water) and 3_{21}-4_{14} (*ortho*-water) in the assumption of rotational equilibrium inside each modification. Both spectral transitions belong to the vibrational ground state of the $H_2{}^{16}O$ molecule.

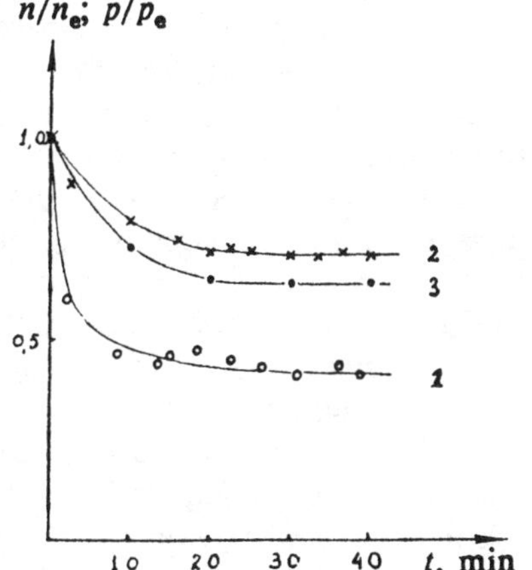

Fig. 1. Bulk concentrations of the spin-modifications of H_2O molecules as functions of time: *1*) *para*-water, *2*) *ortho*-water, *3*) overall water vapor pressure. P_0 is the initial pressure.

Water vapor at a pressure of 3 Torr, produced by vaporizing distilled water, was admitted into both cells, after which the cells were cut off from the water vapor source. Figure 1 shows how the concentration of the modifications varies with time from the instant the water vapor was admitted. The curves are normalized to the equilibrium content of *para*-water 25% and *ortho*-water 75% at 300 K, which was registered immediately after admitting the water vapor. The curves show that the concentration of both modifications decays in the gas phase, which is noted from the drop in water vapor pressure in the cell, however, the decay rate of *para*-water molecules exceeds the pressure decay rate; the situation is the opposite for *ortho*-water. *Para*-water is adsorbed faster than *ortho*-water, so, for example, 10 minutes after admitting, the bulk concentration of *para*-water is 16% and *ortho*-water is 84%. If we continue to observe the *ortho*-*para*-water concentrations, it turns out that the water vapor adsorbent approaches statistical equilibrium with a time constant around 3 hours. The *ortho*-water and *para*-water concentrations in the volume become 3:1, as when they were admitted.

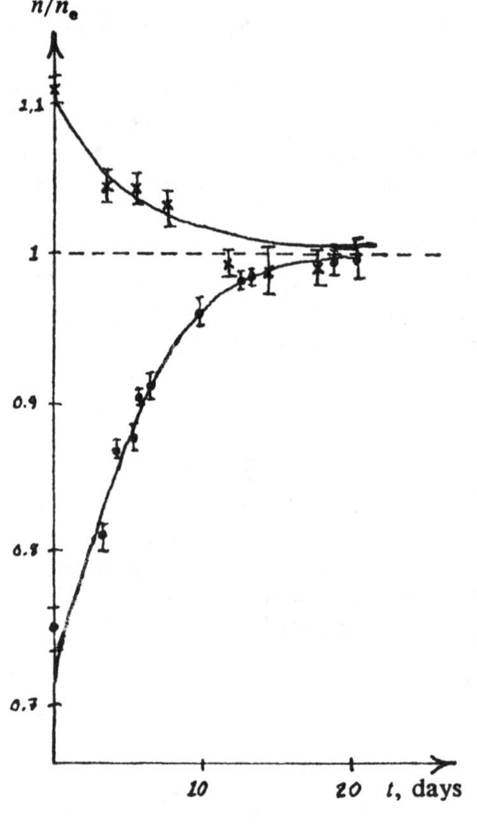

Fig. 2. Relaxation of the water spin-modifications to the equilibrium state: × - *ortho*-water, • - *para*-water (n_e is the equilibrium concentration.

It is reasonable to suggest that the mutual conversion of spin-modifications of water molecules into each other (spin-conversion), as for the spin-conversion of hydrogen, takes place on the surface of a solid, therefore the rate of spin-conversion can depend on the wall material of the cell. To verify this suggestion, an experiment was set up as follows. Ten minutes after admitting water vapor into cells, the nozzle connecting the cells was pinched, the exchange of water vapor between cells ceased, and we continued observing the concentrations of *ortho*-water and *para*-water in the stainless steel cell. Figure 2 shows how the bulk concentrations of *ortho*-water and *para*-water vary in time. The curves are normalized to the equilibrium concentration of the modifications at 300 K. They show that water vapor approaches total statistical thermal equilibrium with a time constant of 4.4 ± 0.2 days, which is significantly longer than in the case when corundum ceramic served as the reaction surface inducing spin-conversion.

The spin-conversion mechanism on the surface remains quite small in comparison with the radiative process both in the experimental and in the theoretical view, therefore it is difficult to predict the conversion rate for specific experimental situations. However, it is possible to define a lower limit to the spin-conversion rate, if thermal dissociation of molecules in the adsorbed state is considered. The dissociation-recombination rate of molecules is indeed the limit, since splitting off a proton destroys the correlation between the proton's spin variables in a water molecule. If we consider that isotopic exchange in the adsorption layer also takes place through molecular dissociation-recombination, then by comparing the spin-conversion and isotopic exchange rates, it is possible to draw a conclusion regarding spin-conversion.

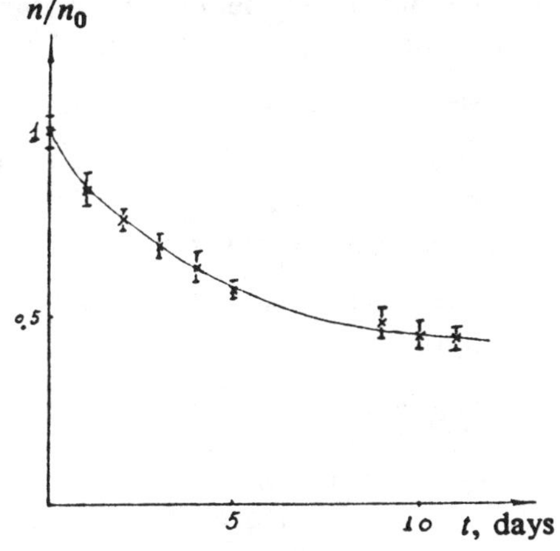

Fig. 3. Bulk concentration of D_2O molecules as functions of time in the process of isotope exchange $H_2O \rightleftarrows D_2O$ on the surface of the stainless steel cell; n_0 is the initial concentration of heavy water vapor.

The experiments on determining the rate of isotopic exchange $H_2O \rightleftarrows D_2O$ on a stainless steel surface were conducted as follows. D_2O vapors were admitted to the same stainless steel cell containing the adsorbed H_2O molecules. The bulk concentration of heavy water was measured as a function of time relative to the total absorption on rotational transitions of molecular D_2O in the submillimeter spectral range. The decrease in molecular D_2O concentration in the gas phase (see Fig. 3) is caused by isotope exchange and the appearance of HDO molecules. The time constant to establish isotopic equilibrium under our conditions was 5 days. The spin-conversion and isotope exchange time constants matching within experimental errors means that there is no specific spin-conversion mechanism for a stainless steel surface at room temperature. The lifetime of nonequilibrium *ortho-para*-water under these conditions is determined by the time molecules exist as a whole.

Experiments on the adsorption of D_2O molecules on a solid carbon dioxide surface were done at liquid nitrogen temperature in order to find the temperature, as the main factor controlling the adsorption process, relative to the possibility of better reproducing the conditions for deposition of D_2O molecules on the surface of fine clusters in a supersonic flow. First, solid carbon dioxide was frozen from the gas phase onto the walls of a glass bulb immersed in liquid nitrogen, then water vapors were admitted to the same bulb. Gas exchange was followed by evacuating the cell to a pressure of 0.1 Torr. After this, the cell was allowed to warm to room temperature, the frozen layer vaporized, and a sample of the gas was taken to the measurement cell of a submillimeter spectrometer. The bulk concentrations of *para*-water and *ortho*-water N_p

and N_o, respectively, were determined from the amount of total absorption on the 1_{01}-1_{10} and 4_{31}-5_{24} transitions using a previously described method [3]. The area of cooled glass surface and the amount of solid carbon dioxide was kept constant from experiment to experiment, and the number of carbon dioxide molecules n_{CO_2} and the number of molecules of deposited water n_{H_2O} were determined from the volume and pressure of the gases. It is further assumed that the composition of the gas after vaporization correctly gives the composition of the solid layer and the water deposited on it, including the content of *para*-water and *ortho*-water.

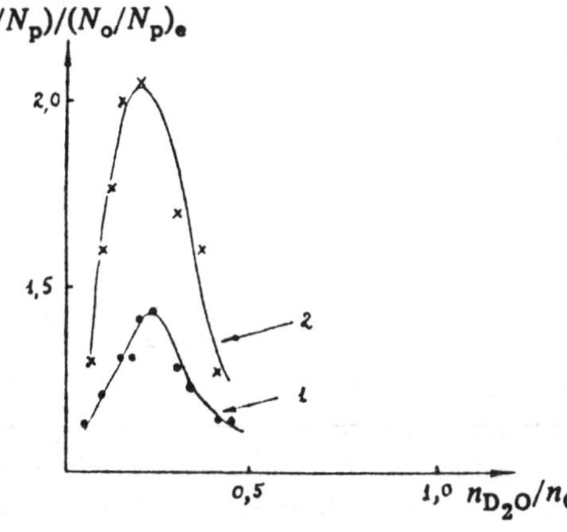

Fig. 4. Spin-modification nonequilibrium of D_2O molecules during adsorption on a surface of solid carbon dioxide as a function of the degree of coating: curve *1*) this work, curve *2*) the results of Ref. [10].

Figure 4 shows the variation (curve *1*) of the relative content of *ortho*-water and *para*-water in the layer as a function of the number of deposited heavy water molecules. The ratio of N_o:N_p is normalized to the equilibrium spin-modification ratio of heavy water (N_o:N_p) = 2:1 at 300 K. The units of the ratio n_{D_2O}/n_{CO_1} on the horizontal axis of Fig. 4 correspond to 22.7 millimoles/gram, as is usually done in works on adsorption.

The deviation of curve *1* from a horizontal line attests to there being *ortho-para*-selectivity during adsorption, and the variation has the form of a curve with a maximum. The selectivity disappears for small and large water coatings of the carbon dioxide layer. Figure 4 also shows curve *2* for comparison, taken from Ref. [10] and plotted in the same coordinates. By n_{CO_2} and n_{D_2O} it is necessary to understand the bulk concentrations of carbon dioxide gas and heavy water vapor on the input to the sonic nozzle. The qualitatively similar paths of the curves are noteworthy.

Ortho-para-selectivity of heterogeneous condensation of D_2O molecules in a supersonic carbon dioxide jet was treated in Ref. [3] as a consequence of the rotational selectivity of the condensation process, since, as had been established, in a bound state molecules undergo transition only from the level $J = 0$, and this rotational level belongs to the *ortho*-modification of heavy water. The rotational selectivity is related to the initial state of condensation, when D_2O-CO_2 dimer nucleation centers are formed. On the other hand, after sampling clusters from the supersonic jet, analysis of the gas mixture (obtained as a result of vaporizing the clusters) showed that there is excess heavy *ortho*-water. This fact was treated as additional proof of the rotationally selective condensation effect.

The qualitative correspondence of curves *1* and *2* prevented us from concluding that the surplus of *ortho*-molecules in the composition of the clusters was only caused by the rotationally selective formation of dimers. The adsorption of water molecules on the surface of fine clusters in the latter stages of condensation should be called to attention, therefore the surplus of *ortho*-molecules can not serve as demonstration of the rotationally selective condensation effect in the stage of nucleation center formation.

The phenomenon of spin-modification selectivity of molecules during adsorption and condensation of gases does not yet have a reliable theoretical explanation. The literature for each case, when selectivity is observed, cites its own explanation, sometimes on the level of hypothesis.

The predominant adsorption of *ortho*-hydrogen at low temperatures is explained by the fact that the adsorption energy of *ortho*-molecules is higher than that for *para*-molecules by the magnitude of a rotational quantum. In the case of the condensation of NH_3 molecules, molecules predominantly condense from the states $J = 1$, $K = 1$, and $J = 2$, $K = 2$, for which the angular momentum is directed along the molecule's symmetry axis. The dipole moment of NH_3 molecules, which is also directed along the symmetry axis, is a maximum for these states, which leads to higher dipole-dipole interaction energy in comparison with other rotational states. An interesting case of selectivity is realized with the formation of H_2-HF dimers in a supersonic gas jet. The low bonding energy of the molecules in the complex and the small mass of the H_2 molecules lead to large amplitude of zeroth-order vibrations about the angle relative to the line connecting them. If before formation of the complex the H_2 molecule is in the state $J = 0$, then the part of the wave function which describes the orientation of the H_2 molecule in the complex turns out to be nearly spherical. If, however, the initial rotational state is $J = 1$, then the H_2 molecule in the bound state turns out to be oriented perpendicular to the line connecting them. Such a T-shaped configuration has a higher bonding energy and, therefore, complexes are formed predominantly of H_2 molecules at the level $J = 1$.

For H_2O and D_2O molecules, a satisfactory explanation for spin-modification selectivity during adsorption and condensation has not yet been found. It is significant that *ortho-para*-selectivity of adsorption takes place on the surface of an ionic crystal and on the surface of molecular crystal, if they are covered by some number of water molecules. If, on the other hand, there is no water on the surface at all, or the layer of molecules is very thick, which in the limit corresponds to a free liquid water surface, then the selectivity disappears. According to our experimental data, the vaporization of water from a free liquid surface at room temperature always gives water vapor that is at equilibrium in terms of spin-modification.

REFERENCES

1. C. M. Cunningham, D. S. Chapin, and H. L. Johnson, "Separation of *ortho*-hydrogen from *para*-hydrogen and of *para*-deuterium from *ortho*-deuterium by preferential adsorption", J. AMER. CHEM. SOC., Vol 10 No 10 pp 2382-2386, 1957.

2. Y. L. Chandler, "The adsorption and *ortho-para*-conversion of hydrogen on diamagnetic solids II: the relative adsorbabilities of *ortho*-hydrogen and *para*-hydrogen", J. PHYS. CHEM., Vol 58 No 1 pp 58-61, 1954.

3. V. K. Konyukhov, A. M. Prokhorov, V. I. Tikhonov, and V. N. Faizulaev, "Effect of rotational selectivity during heterogeneous condensation of water vapor", this volume.

4. H. D. Barth and F. Huisken, "CARS spectroscopy in supersonic jets of ammonia monomers and clusters", J. CHEM. PHYS., Vol 87, No 5 pp 2549-2559, 1987.

5. C. M. Lovejoy, D. D. Nelson Jr., and D. J. Nesbit, "Hindered internal rotation in jet cooled H-HF complexes", J. CHEM. PHYS., Vol 87 No 10 pp 5621-5628, 1987.

6. V. K. Konyukhov, V. I. Tikhonov, and T. L. Tikhonova, "Spin-modification selective adsorption of water molecules on a surface of corundum ceramic", KRAT. SOOBSHCH. FIZ., No 9 pp 12-14, 1988.

7. V. K. Konyukhov, V. I. Tikhonov, and T. L. Tikhonova, "Ortho-para nonequilibrium during the adsorption of water molecules by carbon dioxide surface", KRAT. SOOBSHCH. FIZ., No 12, 1988.

8. B. G. Linsen (editor), "Physical and Chemical Aspects of Adsorbents and Catalysts", London-New York, Academic Press, 1970, 653 pp.

9. J. C. Polanyi and R. J. Wolf, "Dynamics of simple gas-surface interaction II: Rotationally inelastic collisions at rigid and moving surfaces", J. CHEM. PHYS., Vol 82 No 3 pp 1555-1576, 1985.

10. V. K. Konyukhov, V. I. Tikhonov, and T. L. Tikhonova, "Separation of spin-modifications of water and heavy water molecules", PIS'MA ZH. TEKH. FIZ., Vol 12 No 23 pp 1438-1441, 1986.

SUBJECT INDEX